The history of astronomy
from Herschel to Hertzsprung

The history of astronomy from Herschel to Hertzsprung

DIETER B. HERRMANN

Director of the Archenhold Observatory, Berlin-Treptow

TRANSLATED AND REVISED BY
KEVIN KRISCIUNAS

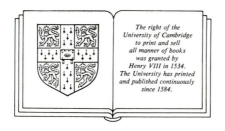

The right of the
University of Cambridge
to print and sell
all manner of books
was granted by
Henry VIII in 1534.
The University has printed
and published continuously
since 1584.

CAMBRIDGE UNIVERSITY PRESS

CAMBRIDGE

LONDON NEW YORK NEW ROCHELLE

MELBOURNE SYDNEY

Published by the Press Syndicate of the University of Cambridge
The Pitt Building, Trumpington Street, Cambridge CB2 1RP
32 East 57th Street, New York, NY 10022, USA
296 Beaconsfield Parade, Middle Park, Melbourne 3206, Australia

Originally published in German as *Geschichte der Astronomie von Herschel bis Hertzsprung* by VEB
Deutscher Verlag der Wissenschaften, 1973 and © VEB Deutscher Verlag der Wissenschaften
1973
Third edition first published in English by Cambridge University Press 1984 as *The history of
astronomy from Herschel to Hertzsprung*
English edition © Cambridge University Press 1984

Printed in Great Britain at the University Press, Cambridge

Library of Congress catalogue card number: 83-15427

British Library cataloguing in publication data
Herrmann, Dieter
The history of astronomy from Herschel to Hertzsprung
1. Astronomy—History
I. Title   II. Geschichte der Astronomie
English
520'.9'03   QB28
ISBN 0 521 25733 6

Contents

Preface to the English edition

The past two hundred years represent one of the most fruitful epochs in the history of astronomy. During this time intensive work by a comparatively small number of astronomers created the essential foundations of today's astronomical world view. The astonishing depths of the universe became accessible to our eyes and minds; the laws of the universe were rigorously pursued by the investigators.

There has been a close mutual relationship between astronomical research and the general development of society, especially at the time that worldwide maritime trade was developing. The determination of longitude and latitude at sea stimulated numerous theoretical and practical investigations of a fundamental nature such as the development of precise catalogs of stellar positions, the theory of the Moon's motion, and the development of more modern measuring instruments, including chronometers. Later the study of the cosmos benefited from the growth of capitalist industry and technology such as the production of greatly improved research instruments.

Having two names – Herschel and Hertzsprung – as cornerstones of our excursion into historical astronomy should in no way imply that science is only the work of individuals. Astronomy, like the other sciences, also grows through collaboration, without which the discoveries associated with famous names would not be possible.

Not only astronomers built the world view of astronomy; rather, the development of cosmic research shows directly that scientific understanding of any complex phenomenon in the universe is possible only with the closest cooperation of different natural sciences. Moreover, astronomy has long brought up questions that belong to the realm of philosophy. It is precisely these far-reaching specific-technical problems that represent astronomy's contribution to our world view.

The following book attempts to trace the development of astronomy through its principal directions or main areas of activity, rather than through the biographies of scientists or a strict chronology of events. Although something is lost thereby in terms of color, this method has the advantage that it clarifies the inner logic of discovery in the different areas of interest. An artificial isolation of areas is avoided because of their close actual interrelationships.

The history of astronomy in the nineteenth and twentieth centuries involves a great number of interrelated, detailed investigations which, in the scheme of the present work, cannot be presented sequentially and to a desirable degree of completeness. The reader who is interested in the historical development of such problems will find in the bibliography a summary of important general and special primary and secondary literature.

The principal purpose of this book is fulfilled if it is useful to all those who are concerned with the development of the modern astronomical world view.

The History of Astronomy from Herschel to Hertzsprung has so far gone through three German-language editions published by the VEB Verlag der Wissenschaften. Comments and suggestions by readers have greatly improved the book. I thank all these attentive readers. Special thanks are due to Prof. Dr habil. F. Herneck (Berlin) for constructive help in the development of the manuscript, as well as to Mrs U. Sell, the Reader of the book at the Verlag der Wissenschaften. I am pleased that Cambridge University Press is publishing an English-language edition of the book. This is the result of extensive efforts on the part of the translator, Mr Kevin Krisciunas (United Kingdom Infrared Telescope Unit, Hilo, Hawaii), who intensively dedicated himself to the manuscript. Finally, I would like to thank Dr Simon Mitton of Cambridge University Press.

D. B. HERRMANN

Berlin (GDR), January 1983

Translator's preface

Since the appearance in 1902 of the fourth edition of *A Popular History of Astronomy during the Nineteenth Century* by Agnes M. Clerke, there has been no other major English-language monograph covering that era of the history of astronomy. The present work, which covers a somewhat longer period of time (from about 1780 to 1930), provides the reader with a shorter, but much more well-rounded, overview of the trends in science and technology which led up to the present century's golden age of astrophysical research. Compared to Miss Clerke, Dr Herrmann has had the advantage of hindsight and therefore has had the option to elaborate on the areas of development which have the greatest relevance to our present-day astronomical concerns, such as galactic astronomy, cosmology, and the never-ending struggle to study celestial objects with more advanced telescopes and by means of more sophisticated methods of data analysis.

The development of astronomy from Herschel to the present day is best exemplified by a shift from positional astronomy to astrophysics; from questions like 'How many stars can we count?' to 'What are the stages of physical and chemical evolution of the stars?' There has been a marked shift from visual observing to photography, photometry, spectroscopy, and complex electronic detectors. As the reader will learn, astronomical research progressed hand in hand with the opportunities provided by the technological advances. In the nineteenth century this meant quality control for such things as lenses and micrometers, and being able to aluminize mirrors. In the twentieth century there came good photographic emulsions, photomultiplier tubes, and diffraction gratings. Now with computer technology and microelectronics we can gather and process more data than ever before.

From the era of Herschel to that of Hertzsprung scientists greatly advanced our state of understanding of the universe. As most astronomy books deal with the era of Copernicus to Galileo or deal only with present-day activities of astronomers, the reader will be pleased that Dr Herrmann has worked on an important period in between, and I am certainly pleased that Cambridge University Press is publishing this translation.

I would like to thank Paul Luther of Bernardston, Massachusetts, for carefully going through the manuscript. He pointed out a number of questionable passages, some of which are in the original, and also pointed out where certain references to other astronomers and their works should be made. Dr Michael Hoskin kindly checked some quoted passages. Karin Fancett did a very thorough job with the editing of the manuscript.

I would also like to thank Mr Ralph Boersma of Warrenville, Illinois, for providing me with a good foundation in the German language; and Dr Harry Haile, Professor of German at the University of Illinois at Urbana-Champaign, for inspiring me to combine my interests in science and language.

The footnotes marked by asterisks have been added by me. They contain supplementary information or clarifications of certain terms or ideas. I have filled out the name index with as many full names and birth and death dates as possible.

As for some food for thought on the art of translating, I could do no better than quote Vladimir Nabokov:

> In the first place, we must dismiss, once and for all the conventional notion that a translation 'should read smoothly', and 'should not sound like a translation' (to quote the would-be compliments, addressed to vague versions, by genteel reviewers who never have and never will read the original texts). In point of fact, any translation that does *not* sound like a translation is bound to be inexact upon inspection; while, on the other hand, the only virtue of a good translation is faithfulness and completeness. Whether it reads smoothly or not, depends on the model, not on the mimic.*

Though there are obvious differences between translating a classic of Russian literature and translating a history of astronomy, trying to please one's readers while remaining true to the original text is a perennial problem. My first version of this book was overly literal, and though Nabokov might have been pleased with it, making it less literal has made it better. I hope that the present result will be informative in a style that is not too stiff. While I was working on this translation I accumulated quite a correspondence with Dr Herrmann regarding specific facts, phrasing of certain passages, and the contents of the appendices. A number of paragraphs have been eliminated from the original (the third German edition) with the consent of the author, but additional material specifically on the Marxist attitude toward evolution may be required for an understanding of the section 'Evolution in the universe'. Elsewhere in the book I, in my role as 'mimic' have taken the liberty of rephrasing certain sentences, since I felt that the 'model' could not be clearly conveyed however it was translated.

KEVIN KRISCIUNAS
Hilo, Hawaii, 13 March 1983

* From the foreword to Nabokov's translation of Mihail Lermontov's *A Hero of Our Time* (Garden City, New York: Doubleday), 1958, p. xii.

Introduction

Modern astronomy begins with the scientific work of Nicholas Copernicus, whose *De Revolutionibus Orbium Coelestium* (On the Revolutions of the Heavenly Spheres) was published in 1543 in Nuremberg. Without the bold hypothesis set forth by the Polish scholar in this book the further development of astronomy would not have been possible.

The achievement of Nicholas Copernicus was that he overturned the astronomical world view handed down from the ancients, in which the Earth is the center of the universe, and he explained that the Earth was to be considered a planet just like the other wandering stars that revolve around the Sun, which is placed in the center of the universe. Copernicus explained the movement of the heavenly bodies around the Earth as only apparent; in actuality it is the reflection of the rotation of the Earth about its axis. According to Copernicus the complicated motions of the planets in the sky come about through their actual motion around the Sun coupled with the simultaneous motion of the Earth. As a result Copernicus had not only replaced one hypothesis with another that seemed to him simpler and more harmonious, but he had explained the essentially correct structure of the planetary system.

Copernicus could not prove his daring assertion conclusively, and his revolutionary book contains many leftover elements of ancient Ptolemaic astronomy. For example, Copernicus took great care to describe all motions of heavenly bodies as circles, in order to satisfy a demand required by the ancients for geometrically ideal orbits of heavenly bodies. The most serious consequence of this scientifically unjustified restriction on the theory was that Copernicus had to resort to the geometrical tools of ancient astronomy.* As a result the advantage of clarity and simplicity, which would possibly convinced many astronomers, was still absent. The doubt of the scientific world was directed also to the immediately failing proof for the hypothesis. If the Earth, like other planets, actually revolved around the Sun, then this motion must necessarily be reflected in an apparent motion of the fixed stars in the sky. In fact the astronomers during

* Copernicus' use of ancient geometrical tools seems 'unjustified' only in the context of our *modern* scientific methods. (Tr. note)

I

the time of Copernicus could not detect such periodic shifts of the fixed stars. This subsequently led to one of the most fruitful contradictions in the history of astronomy. For the proponents of the Copernican system and for Copernicus himself the inability to measure parallaxes was by no means considered a proof against the motion of the Earth about the Sun. He saw in this an expression of the undemonstrated great distances of the stars, which he thought were placed on a sphere. The necessity of continuously improving the precision of measurements was the result of this. If the fixed stars were not situated too far away in space, they must show a parallax, and the proof of this was finally only a question of the precision of the angular measurements. The continual efforts for the development of precise measurements, which were later stimulated through other motives, went on for almost three hundred years before the demonstration of the first stellar parallaxes (see pp. 45 ff.).

The great German astronomer Johannes Kepler made substantial improvements to astronomy immediately after Copernicus. From the beginning of his career he had believed in the heliocentrism of Copernicus. This faith guided him through his decades of research into the interpretation of valuable observations of the Danish astronomer Tycho Brahe and led to an important system of knowledge of the planetary system. In his books *Astronomia Nova* (1609) and the later work *Harmonice Mundi* (The Harmony of the World, 1619) he developed three laws in mathematical form describing the motion of the planets about the Sun. With these laws Kepler proved for the first time the existence of a strict tendency toward natural law and, moreover, in his *Tabulae Rudolphinae* (Rudolphine Tables, 1627), based on these laws, he created the foundation for the calculations of planetary positions with unprecedented accuracy. At the same time Kepler corrected the Copernican world view on a basic point. He proved that the planets do not move in circles but in elliptical orbits around the Sun. This correction demonstrated that nature is not subject to the commands of ancient authorities; nature can be known only through observations and their theoretical analysis.

The existence of so strict a law led the ever-pondering Kepler to the question of what would be the source of the effect of this law. After much consideration he arrived at the conclusion that the Sun must be the place of force, a phenomenon he regarded as similar to magnetism. Kepler did not further succeed in promoting the transition from the kinematics to the dynamics of planetary motion, but his work led up to the threshold of celestial mechanics.

The English scholar Isaac Newton made the decisive transition to the founding of celestial dynamics in his fundamental work *Philosophiae Naturalis Principia Mathematica* (Mathematical Principles of Natural Philosophy, 1687). In it Newton proved that a force of attraction operates between two masses which is proportional to the product of the two masses and inversely proportional to the square of their separation. The three Keplerian laws of planetary motion may themselves be derived from this gravitational law. The source for the behavior of the planets, formulated by Kepler, was found in the gravitational force.

The discovery of the law of general attraction of masses was a milestone in the history of science. Copernicus had already demonstrated the difference between heaven and Earth, as he made the Earth essentially equal to the other planets, but Newton then extended the range of this equality further. The same force which causes the free fall of bodies on the Earth (the law which the Italian Galileo had investigated) is also the source of planetary motion. Galileo's experiments played an important role in the investigation of the Law of Gravity. Newton was able to show that the Moon is just like a stone which 'falls' around the Earth.

Three quarters of a century before the appearance of Newton's work an epoch-making invention became known – the astronomical telescope. The telescope strengthened the principle of the similarity of heaven and Earth, a principle which various ancient thinkers had wholeheartedly rejected. Dutch lens grinders had accidentally discovered the magnifying effect of certain combined optical glasses, and during the year 1610 Galileo became the first educated person to comprehensively make use of this new optical device which multiplies the power of the human eye. Within a few weeks he discovered numerous phenomena which testified against the Aristotelian–Scholastic interpretation of the heavens. Kepler praised the new means of research as a 'much-knowing perspicil, more valuable than any scepter'.[1] Galileo discovered the crescent phases of Venus; was the first to see the mountain world of the Moon which looked like a second Earth; revealed four bright small stars next to Jupiter which revolved about this planet like the planets around the Sun according to Copernicus; and lastly resolved the diffuse light band of the Milky Way into innumerable tiny points of light. Although these facts did not allow the new world system of Copernicus to be proven, they fortified Galileo and many contemporaries in their conviction of the correctness of the heliocentric world view.

The breathtaking chain of great events which sparked the systematic growth of the Copernican system arose under the strict social conditions of the Renaissance with its exasperating conflicts and struggles, and was itself an important part of this struggle.

The reordering of the cosmos by Copernicus did not concern itself only with astronomical knowledge. Viewed in ideological hindsight, the pioneering book *De Revolutionibus* contained more explosive material than its creator could have perceived. Of greatest significance was the philosophical consequence of the oneness of the Earth and cosmos, from which Copernicus declared the Earth to be a planet essentially like the other planets, which were also heavenly bodies. Herein lies one of the kernels of thought of the material nature of the universe. The subsequent possibility of developing

celestial mechanics on the basis of the Copernican world view rests upon
this, while at the same time materialism in general helped to develop
mechanics as a science.

Also important for the further development of the natural sciences was
the distinction made between actuality and appearance emphasized by
Copernicus, and the knowledge connected with it that the actuality of a
thing does not immediately result from appearance. Research must primarily advance with respect to reality.

The origin of the heliocentric representation, its development, and its
consequences are closely connected with revolutions which took place in the
realm of society, in the economics and politics of the Renaissance. The work
of Copernicus most deeply shook the authority of the medieval Church,
whose teaching was an undeniable part of the ideology of feudalism. The
system of Copernicus not only stood in contradiction to Christian teaching
but also directly touched upon the ruling power structure with its challenge
for a previously unknown freedom of thought. Goethe wrote: 'Perhaps a
greater challenge to mankind has not taken place, for everything vanished
into vapour and smoke with this acknowledgement: a second Paradise, a
world without guilt, [a world of] poetry and piety, the testimony of the
senses, the conviction of a poetically religious belief.'[2] All this was not
clearly discernible. The relationship of the Church to the work of Copernicus was not hostile from the start; the Church was very interested in
fundamental astronomical studies, as a defective knowledge of the motions
of the stars caused the calendar to 'run false' by ten days. Most importantly,
this touched upon the determination of Church holidays. An astronomical
system which promised to bring order to this chaos could only be welcomed
by the Church. Nothing is known of opposition of the Church to the ideas of
Copernicus circulated in the handwritten *Commentariolus*.

With the work of Copernicus science again objectively repeated the
burning of the Papal Bull by Martin Luther, as Engels formulates, and had
thrown down 'the gauntlet of ecclesiastical authority in natural things'.[3]
When subsequently the ideological consequences of the new teaching
became clearer, the Church put on this 'gauntlet' and mustered all its
strength against the dissemination of this new world view. In 1600 the
Italian scientist, philosopher, and avowed Copernican Giordano Bruno
paid with his life for his daring thoughts on the multiplicity of inhabited
worlds. In 1616 the work of Copernicus was placed on the Index of
Prohibited Books of the Catholic Church. Galileo was dragged before a
tribunal of the Inquisition in 1616 and 1632 for his acceptance of the new
world system. After gruelling trials he was finally forced under threat of
torture to openly renounce his conviction and to give up all arguments for
the new world view. These campaigns against specific parts of the natural
sciences and their operations failed however, because natural science was
not an isolated subject which willfully allowed itself to be silenced; it was a
component of an all-embracing development.

In the middle of feudalism a new class arose and strengthened itself – the bourgeoisie. This class grew and spread, and with this growth feudalism was liquidated. The bourgeoisie had a principally different attitude toward the natural sciences. The bourgeoisie built the historical stage on which production and the whole system of economic relations changed. A new social class was built in the cities by manual laborers and merchants through extensive industrial production. A new climate arose for science. On the one hand the inventions and technical achievements of trade were a source of stimulation for experimental work which played a corresponding role in the renewal of natural sciences; on the other hand commerce produced the market, and this fostered trade. The striving for replacement of usable land routes to the traditional places of commerce and to previously unknown territories by new, cheaper sea routes could not have been realized without the help of science. The most immediate and historically earliest contact between science and the spread of specific results of early capitalist production occurred at this time. The single science upon which it depended was astronomy. For his discovery of America Columbus used the astronomical tables of the German astronomer Regiomontanus, a scholar who had expressed cautious doubt in his investigations concerning the correctness of the Ptolemaic system. It was completely immaterial for the development of sea commerce, which eventually changed the whole economic equilibrium in favor of seafaring nations, which place in the cosmos the Earth and Sun took. But the accurate calculation of ephemerides was closely connected with the perception of the truth concerning the structure of the planetary system. This is not a simple process and there is no reason to suppose that the scholars who participated in this process knew their particular historical roles, but the close connection cannot be overlooked. The further growth of the bourgeoisie, and with it capitalism, confirmed this. England provided an impressive example of this close mutual dependence where scientific life flourished like nowhere else on Earth after the ascendancy of Cromwell (1649). The industrial revolution in England which subsequently spread is also to be understood on this basis. It exerted again an extremely demanding pressure on the sciences, among them astronomy. For the fulfillment of important precision tasks astronomy even at this time needed some kind of instrumentation works. This could only develop and operate where machine production, metallurgy, and the technical spirit of invention had reached a high level of development. However, the sciences still profited more from this technical development than the reverse. More and more, with its methods of geographical determinations of position, astronomy provided a solitary contribution to the solution of questions which stood in direct relation to the development of the bourgeoisie and capitalism.

The controversies concerning the astronomical world view were in no way at an end. Rather, it happened that the bourgeoisie, which had arisen in the struggle against feudalism, stressing antireligious feeling, needed religion itself for the maintenance of its power as a 'moral foundation of the state'. The relatively clear attitude in favour of natural scientific results and against the Church remained of interest to the bourgeoisie only so long as it furthered the decline of ecclesiastical power. Once the goal was achieved, the attitude of the bourgeoisie became necessarily disunited concerning the sciences. In one way this was increasingly a condition of the development of capitalism; in another the natural sciences always produced new antireligious arguments of the greatest weight.* When principal religious dogmas were touched upon, bourgeois philosophers always reacted with sharp attacks and were not deterred by misrepresentations. One measure of the position of different philosophical trends toward the natural sciences, which has remained a sensitive issue up to modern times, is the attitude toward the theory of evolution. The pioneering breakthroughs concerning evolutionary processes in nature and society which have prevailed since the eighteenth century are always subjected to bitter attacks by bourgeois philosophers. One need only think of the controversy concerning the theories of Charles Darwin, Ernst Haeckel, or modern cosmology.

---

* This would seem to imply that scientists then and now are to be considered antireligious – an unfounded idea. For example, Galileo, Kepler, and Einstein were able to retain deep religious beliefs. (Tr. note)

I

Construction and motion of the heavens – classical astronomy

ON THE CONSTRUCTION OF THE HEAVENS*

The remarkable progress which astronomy made toward the end of the eighteenth century is associated with several factors found in the state of development of astronomy itself as well as in the meaning of astronomy for society. Practical research, new ideas and developments, new technical possibilities – all these led to a marked acceleration of the tempo of development of astronomical research.

The discovery of a new large planet of the solar system is one of the developments which restimulated the interest in particular questions. This planet was accidentally found on the night of 13 March 1781 in the course of a stellar survey by William Herschel, who came from Hanover, Germany, and was active in England after 1759. In the course of his research – to measure stellar parallaxes of large magnitude – Herschel discovered an object in the constellation Gemini which exhibited a greater image size in comparison to the fixed stars. Herschel first thought it to be a comet. Later it was shown that it undoubtedly had to be a planet – Uranus was discovered. The impact of this discovery on the public was considerable; for the first time since the beginnings of astronomy in the distant past a new planet had been found. All other planets had been known since the very beginning. The new heavenly body was situated almost twice as far from the Sun as Saturn. New distances were presented to the eyes of the world. More significant than the discovery itself, which would have been made sooner or later by other astronomers,[1] was that Herschel, encouraged by this success, resolved to dedicate himself exclusively to astronomy and to give up his regular position as a musician. Herschel was one of the exceptional research personalities of the turn of the nineteenth century. His works on almost all areas of astronomy correspondingly influenced the further progress of research. He studied the planetary surfaces with great success, dedicated himself to comets and the Sun, and was at the same time an ingenious telescope builder (see pp. 159 ff.). However, his bold, pioneering ideas and

---

* The term 'construction of the heavens', though it sounds somewhat awkward, is Herschel's own (in English), so we shall leave it as it is. (Tr. note)

Fig. 1. William Herschel

research on the 'construction of the heavens' and the developmental processes in the universe were most significant for the development of astronomy.*

* William Herschel would not have been able to accomplish as much as he did without the lifelong assistance of his sister Caroline. Not only was she often on hand to copy down her brother's observations while he was glued to the telescope, but she herself independently discovered eight comets and three nebulae (one being a companion of the Andromeda Galaxy), and her 1798 revision of Flamsteed's catalog (see p. 30) was published by the Royal Society. (Tr. note)

At that time wondering about the arrangement of stars in space involved nothing self-evident. For Copernicus the fixed stars were still thought of as arranged on a sphere, about which one could say nothing other than it must be very distant. Even Kepler still considered the fixed stars as objects on a thin spherical shell.

The astronomical telescope had already allowed a previous conjecture to be proven which was brought into the new investigation; the idea that the Milky Way consists of single stars also deserves to be rated highly among Galileo's telescopic discoveries. As a result the question necessarily arose as to why so many stars are found in a great circle which circumscribes the whole heavens. This circle is represented by the band of the Milky Way and is not seen in other places of the sky.

After there was an understanding of the spatial structure of the planetary system there arose the question of the structure of the cosmic surroundings of the planetary system.

The first ideas which touched upon the distribution of stars in space were published by the Englishman Thomas Wright in 1750. Here for the first time the thought was expressed that the impression which we get from the distribution of stars in the sky depends in part on the specific spatial distribution of the stars as well as the vantage point on the Earth as an island in the cosmos. According to Thomas Wright the Sun with the planets surrounding it should find itself in an extensive layer of stars which has a relatively minute thickness. An observer who finds himself inside this layer must then actually perceive the stars along a great circle in particularly large numbers. If the observer looks perpendicular to this plane in the sky, a substantially smaller number of stars would be noticed because of the minute thickness.

This work by Wright contributed to the appearance of another significant work: *Allgemeine Naturgeschichte und Theorie des Himmels* (General Natural History and Theory of the Heavens, 1755), a genial early work by the German philosopher Immanuel Kant. In this text Kant expresses himself in the same vein as Wright concerning the phenomenon of the Milky Way: 'The form of the starry heavens has thus no other explanation than just such a systematic constitution on the whole that the planetary system has in miniature, in which all stars make up a system whose general plane relation is the Milky Way.'[2]

The views of Thomas Wright and Immanuel Kant on the distribution of stars did not originate from analyses of extensive observational material. The German scholar J. H. Lambert also arrived at a similar view and addressed himself to the question of the spatial arrangement of the stars in his *Cosmologische Briefe* (Cosmological Letters, 1761). For him there were systems of different order in the cosmos which exhibited their structure according to significant similarities. From the existence of the planets with their moons and the Sun with its planets Lambert reasoned analogously that the stars were situated in giant spherical star clusters which themselves make up an even larger system. Finally, all of these clusters should be situated in a region of comparatively small thickness. Lambert thought it possible as a basis of his analogies that more extensive systems of greater magnitude also existed, and they all possessed a central body like the planets or the Moon. All these works have their common speculative bases; thus it was that William Herschel did a great service by having begun systematic empirical studies on these questions. His numerous treatises on the structure of the heavens, which he began in 1784, became permanent monuments of science.

Herschel then attempted to put together a picture of the distribution of stars in the sky as accurately as possible, basing his picture on the already existing qualitative knowledge. But the data of later positional catalogs were not placed at his disposal. The data which he could fall back on could not be satisfactorily used for the intended purpose, so Herschel undertook an extensive, accurate survey of the heavens, for which his 20-foot reflector was to be used. The field of view of his instrument had a diameter of 15 minutes of arc (about half the Moon's diameter). It was not possible to completely survey the whole sky with this telescope; therefore, Herschel restricted his counts to a total of 3400 fields, and even this was an extremely toilsome and arduous undertaking (Fig. 2). In the densely stellar regions of the sky it required the enumeration of a nearly unresolvable swarm of light points. In the constellation of Cygnus alone there were tens of thousands of stars visible in the field of view of the stationary telescope over the course of three-quarters of an hour.

The laborious work of these star gauges – as Herschel called his procedure – paid off: if one supposed, as did Herschel, that in every small region of the universe there is on average the same number of stars, then one can determine from the counts the spatial arrangement of the system in which the fixed stars are bound up. An observer whose gaze is directed through a telescope into the universe sees with it a conical volume. The vertex of this cone lies at the focus of the telescope, and its base lies at the ultimate distance which the telescope reaches. The volume of this cone increases with the third power of the distance, as do the number of stars appearing in the field of view. From the respective star counts Herschel determined the relative distances and in this way constructed a picture of the spatial distribution of stars (Fig. 3), which on the whole corresponds to the facts and principally agrees with the representations developed by Kant and Lambert. Herschel thus determined that the Sun takes up its position not far from the centre of the starry island, a result which much later (1918) proved to be wrong. In any case Herschel confirmed that the Sun was one of many other suns which are jointly situated in the plane of a layer of stars.

394

Gerade Aufsteig.		Polar-Distanz.		Stern-menge.	Fel-der.	Gerade Aufsteig.		Polar-Distanz.		Stern-menge.	Fel-der.
22	4	62	7	48.		50	16	60	55	186.	½
22	4	56	16	39, 6	5	51	4	57	26	84.	
22	19	104	6	14.		51	32	108	26	36, 8	5
22	37	103	45	30.		52	49	115	30	26, 2	5
24	3	115	10	35.		54	4	57	18	93.	
24	4	109	35	35.		54	8	91	14	328.	¼½
24	7	102	31	30.		54	55	104	23	180.	
24	10	92	59	88.		55	4	108	41	80.	
24	43	103	39	25.		55	16	62	31	206.	¼½
25	37	102	34	39.		59	8	91	14	228.	
26	17	98	8	111.		59	26	72	37	40.	
26	25	103	57	60.		XIX 1	2	71	40	75.	
26	47	97	43	250.		1	34	56	47	127.	
27	1	120	58	30.		2	29	74	53	204.	¼½
27	55	120	44	32.		2	37	103	16	160.	
28	7	102	51	13.		2	49	121	39	14, 1	
28	8	91	44	39.		3	34	55	56	146.	¼½
28	25	103	24	20.		6	34	61	8	196.	¼½
28	37	122	25	12.		7	34	56	56	130.	¼½
29	25	103	24	20.		7	52	57	59	116.	¼½
29	47	97	50	150.		8	38	92	8	120.	
29	49	121	39	24.		9	37	109	1	60.	
30	34	57	18	62.		9	40	56	51	130.	*
31	10	92	42	13, 7	7	12	59	75	21	58.	
31	10	108	53	74.		13	50	59	59	256.	¼½
31	18	103	19	112.		13	52	59	29	158.	
31	17	97	53	188.	½*	14	2	72	15	60.	
31	34	62	34	76.		14	4	61	21	279.	⅓
31	49	121	39	19, 3		14	55	103	36	64.	
33	4	108	43	88.		15	40	55	26	160.	
33	7	103	53	146.	½	16	50	60	43	296.	¼*
34	5	98	34	130.		16	59	73	23	56.	
34	47	71	53	78.	*	17	44	108	12	50.	
34	58	60	41	80.		18	23	78	9	196.	¼⅓½*
36	34	110	12	83.		18	28	61	21	279.	
36	34	91	37	176.	¼½*	19	52	57	14	180.	
36	47	72	28	224.	¼½	19	56	108	12	55.	
37	34	93	29	5.		20	51	60	55	384.	¼¼½*
38	1	104	14	118.	½¼	21	1	78	47	472.	¼½
39	40	93	52	116.		21	34	55	17	208.	
40	28	92	47	10.		22	27	62	29	320.	
40	47	71	48	236.	¼*	24	36	56	49	224.	¼
41	22	91	37	156.	¼	24	49	104	24	36.	
42	49	121	39	15, 2		24	50	60	43	296.	¼
43	17	72	8	368.	¼*	24	53	113	51	18, 3	
43	33	119	21	21.		25	4	57	9	190.	¼½¼
44	34	112	43	83.		25	16	64	18	280.	¼
44	34	60	34	84.		25	22	59	36	340.	
47	32	91	14	328.		25	37	103	50	55.	
48	4	110	12	83.		27	36	72	34	424.	¼*

Fig. 2. Section from the observing tables for Herschel's star gauges

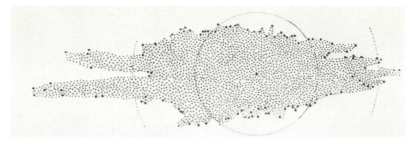

Fig. 3. Structure of the Milky Way system according to William Herschel

Terrestrial observers see this

and, calling this a sidereal stratum, an eye placed somewhere within it will see all the stars in the direction of the planes of the stratum projected into a great circle, which will appear lucid on account of the accumulation of the stars; while the rest of the heavens, at the sides, will only seem to be scattered over with constellations, more or less crowded, according to the distance of the planes or number of stars contained in the thickness or sides of the stratum.[3]

In spite of this Herschel was still not satisfied with the result because nothing could be said about the dimensions of the stellar system. A useful idea occurred to him to combine the luminosities of the stars with the star gauges, and with this method he became the founder of stellar statistics. Twentieth-century astronomy has addressed itself to the questions posed by him with completely similar methods for the progressive state of knowledge. Herschel deduced that information concerning the distances of the stars should be contained in the apparent magnitudes. 'The statement, however, that one with another the faintest stars are at the greatest distance from us, seems to me so forcible', he wrote, 'that I believe it may serve for the foundation of an experimental investigation.'[4] In order to be able to measure the luminosity with sufficient accuracy he discovered the 'principle of light equality': the different stars are observed with two reflecting telescopes, from which it has been proven that they portray a particular star equally as bright. By actual stopping down of a reflector Herschel then sought stars whose magnitude agreed with another in the wide open reflector. Thus he could carry out intensity determinations. If, for example, there was a mirror with half the diameter of another, i.e., stopped down to one-quarter the surface area, then a star which is observed to have equal luminosity in this instrument as in the other, open reflector must have a light intensity four times greater. This photometric procedure was much more useful than the mere counting of stars.

Herschel estimated the dimensions of the Milky Way system from these extensive data. For his estimate he used the distance to Sirius as the unit of distance, because absolute distances to stars were still unknown. According to his research the system of fixed stars to which the Sun belongs has a

Fig. 4. Small Herschelian reflector with wooden tube. Below left the metal mirror made by Herschel

diameter of 850 Sirius distances and a thickness of 155 Sirius distances. Thus Herschel became the first observational astronomer to make a survey of the distribution of stars in space, though he was himself quite certain that the assumptions from which he worked could not unconditionally be true. A star of the next brighter magnitude class must not necessarily have twice the light intensity. Scientific photometry of the heavens later concerned itself with this supposition and proved another relationship between magnitude classes and the light intensities (see pp. 75 ff.). The supposition that all stars exhibit the same absolute luminosity is likewise only fulfilled to a very crude approximation. Because no stellar parallaxes were known, this matter could not be decided. Finally, Herschel had discovered a number of phenomena with his celestial surveys which allowed him to surmise that a great part of the universe is filled with a light-absorbing material; however, here too he could not be absolutely certain. The existence of such a material found between the stars must likewise have a systematic biasing effect on the distance determinations according to the photometric method.

In spite of this drawback it remains an historical triumph of Herschel to have developed for the first time the method for determining the structure

of the Milky Way system. Herschel also knew that a lot was still left to be
done in this area:

And though my single endeavours should not succeed in a work that seems to
require the joint effort of every astronomer, yet so much may we venture to hope,
that, by applying ourselves with all our powers to the improvement of telescopes,
which I look upon as yet in their infant state, and turning them with assiduity to the
study of the heavens, we shall in time obtain some faint knowledge of, and perhaps
be able partly to delineate, *the Interior Construction of the Universe.*[5]

During the entire nineteenth century there were only a few astronomers
who followed in Herschel's footsteps. His son John Herschel began a direct
continuation of the star gauges for the southern sky. The Russian
astronomer Wilhelm Struve undertook numerous investigations concerning the structure of the heavens according to William Herschel's method,
but on the basis of better observational material.* Among other things
Struve discovered that the number of fainter stars increases more slowly
with decreasing angular distance from the galactic plane than the number
of brighter stars. From this he also demonstrated the existence of diffuse gas
and dust between the stars.

However, a more extensive continuation of Herschel's research was not
begun until the end of the nineteenth century.

EVOLUTION IN THE UNIVERSE

In spite of all advances which were realized by natural science after the
Renaissance, and in spite of the great number of newly discovered facts,
science remained prejudiced in one great respect – it regarded all
phenomena of nature as events that happened once and were then
unchangeable. This held for biology, as well as for chemistry, geology, and
astronomy. 'The planets and their satellites, once set in motion by some
secret "first cause", revolved around and around in their prescribed ellipses
for all eternity or at least until the end of all things. The stars are set forever
motionless in their places.' Thus wrote Friedrich Engels.[6]

Even so the intensive and progressive study of activity in nature soon
brought phenomena to light in different sciences, which cast doubt on the
correctness of this metaphysical viewpoint. In astronomy the appearance of
a new star, which abruptly blazed forth in the heavens in the late autumn of

---

* When discussing the work of Wilhelm Struve, one should also mention his son Otto Wilhelm
Struve. Their work closely parallels the work of William Herschel and his son John, in that
the sons very much followed in their fathers' footsteps. And if the Struves were not well
known in their own right, they would be known as the Russian Herschels. As the reader will
soon learn, Wilhelm Struve continued William Herschel's researches in the areas of galactic
astronomy and double stars. Otto Wilhelm Struve continued his father's geodetic work,
determined a value of the constant of precession which was used for half a century, and made
many thousands of micrometric observations of double stars over a period of 53 years. (Tr.
note)

1572, excited general attention and profound considerations by astronomers. As Tycho Brahe concluded by the failure of a measurable parallax, this star must reside in the region which corresponds to Aristotle's unchanging firmament. When in 1604 another new star appeared, it was the opinion of one of its most famous observers, Johannes Kepler, that the star could be the result of the viable potentialities of nature. One could even try to explain its appearance by natural means before one resorted to a creation by God. Otherwise all scientific discussion would stop.

Other facts were no less incompatible with the incorruptibility of the heavens. The long-known comets appeared from nothing and after a short period of visibility disappeared again into nothing. These objects which change their brightness within a short time must have been especially alarming to the metaphysicians. All these things were observational hints which allowed one to conclude that changes also take place in the Aristotelian supralunar region. Immanuel Kant set forth the first coherent attack against the 'strictly fixed view of nature' of the incorruptibility of the heavens, in which he relied on the work of the Frenchmen Descartes and Buffon. On the basis of the Newtonian Law of Gravity, Kant explained the viewpoint in his already-mentioned early work *General Natural History and Theory of the Heavens*, that the planets and their moons, as well as the distant stars and stellar systems, arose from a chaotic primal nebula through a natural developmental process. Kant thereby traced out for the first time an evolutionary theory of cosmic bodies, which consistently refutes a Divine act of creation. As a result he deserves the historically important credit of having introduced evolutionary thoughts to the newer natural science on established grounds. Starting with Copernicus, Kant represents the beginning of the chains of thought which lead, during the nineteenth century, to the philosophical generalizations of the evolutionary concept in the materialistic dialectic of Karl Marx and Friedrich Engels.

Ideologically most significant was the fact that Kant's *Natural History* presented evolutionary events as the effect of natural laws, the prototype of which Kant considered the mechanical law. Thus evolution was regarded by him to be a self-development of the material. Kant expressed this fact in the bold words: 'Give me matter, and I shall build a universe from it.'[7] The *Natural History* was highly praised by the classical writers of Marxism, though it has been relegated along with other so-called early bourgeois accounts of philosophical history to the fringes of Kant's life work. Engels praised the *Natural History* as an 'attack on the eternity of the solar system'[8] and designated it as the 'source of all further progress'.[9] The whole development of natural science has confirmed this assessment.

However, other 'attacks' preceded Kant's 'attack', and subsequent ones followed. The natural significance of fossils established evolutionary thinking in geology, where the work of N. Stensen, R. Hooke, and G. W. Leibniz particularly stand out. Evolutionary thought broke ground in the consideration of the ways of the plant world; Charles Darwin later followed with his theory of evolution in biology. The consequences of these bold thoughts in the realm of geology and astronomy could not be ignored – they required time periods far longer than those calculated in the Bible. Nevertheless, Kepler accepted the reckoning of the age of the world as calculated from Biblical history. According to him the year 1595 was identical with the year 5572 since the creation of the world. Kant, however, needed 'whole mountains of millions of centuries'[10] for the origin of the stellar systems and was thereby in great contradiction to the Bible; the great advocate of the German Enlightenment, Kant's contemporary, G. C. Lichtenberg, ventured to say that the Bible, like other books, was also written by people, and therefore open to interpretation.

The discovery of the structure of the cosmos was very closely associated with a scientific evolutionary theory of the cosmic objects, for the advocates of evolutionary thought had no abstract purpose. Kant and others that followed sought to clarify the origins of the existing structures.

So it is not by chance that structure and development were investigated by Kant in one work according to closely mutual reference. And it is only consistent that the first empiricist of the structure of the universe, William Herschel, simultaneously appeared as the first empiricist of evolutionary thought in astronomy.

Up till this time the subject had been primarily in the domain of natural philosophy. It had attempted to provide whole pictures of nature or single regions of nature in which it 'replaced the still unknown actual relationships with ideal, fantastic ones and supplemented the lacking facts with thought-images'.[11] Herschel's value in this area was that he was the first astronomer to attempt to substantiate these thought-images on the basis of observational materials. He is to be counted, therefore, among those progressive scholars who brought Kant's hypothesis gradually down to Earth.

Herschel began the series of his treatises concerning the construction of the heavens in 1784 and still concerned himself with the development of cosmic bodies in one of his last works (1811). The material which was at his disposal for this was exceptionally voluminous compared to the sparse data with which his predecessors had to work. The peculiar, non-stellar objects, which one calls in astronomy 'nebulae', were hardly cataloged before Herschel. The catalog compiled by the Frenchman Charles Messier consisted of only 103 such objects. By 1802, after Herschel had set up his giant telescope for the systematic survey of the heavens for such nebulae, he enumerated the astonishing sum of 2508 objects in his records. He thereby made the statement that the nebulous objects present themselves in an extraordinarily impressive variety of forms. Little was said, or could then be said, of the characteristics of the different nebulae. A portion were certainly stellar associations. Another portion were definitely not made up of stars; they were formless, concentrated material of whose nature nothing more

could be added. Herschel did not content himself with this general statement. He attempted to classify the objects according to morphological
characteristics, a procedure that also found application in the contemporary
biology.

Herschel altered the classification many times, and it was always the goal
to follow an ordering corresponding to a possible natural evolutionary
sequence of objects. He was firmly convinced that the different objects did
not simply concern representations of a *scala naturae*, an ordering of the
bodies according to a ranking of their complexity, thus a reflection of the
natural hierarchy: it concerned the phases of the evolutionary processes.
While the *scala naturae* represents a static ordering which expresses the plan
of creation, Herschel assumed that a temporal succession of things corresponds to a spatial separation of different objects. He thereby introduced
into astronomy the fundamental principle of every scientific cosmogony,
wherein the cosmic time scale to a certain extent is cut into a small slice; the
subsequent development of a celestial body cannot be observed owing to the
very slow changes going on. He formulated the principle through analogy
to that of biological growth. Indeed, it would almost be the same thing
whether we saw, one after each other during its life, the sprouting,
blooming, breaking into leaf, fruit bearing, withering, drying up, and
decaying of a plant, or if simultaneously one had a great number of
examples in view which correspond to the respective stages of the plant's
life.

In the catalog of nebulae published in 1802 Herschel divided the
observed objects into twelve classes. At the beginning were the single stars;
then the double and multiple stars; the stellar groups according to the
degree of concentration; nebulae considered as possibly made up of stars;
and finally the planetary nebulae, which today still carry this name coined
by Herschel. Herschel interpreted this sequence by means of the Law of
Gravity, whose universal application he presumed, just like Kant, as stages
of progressive evolution. The single stars bind themselves together through
the action of gravity, first with a few other stars into loose groups, and later
into star clusters with increasing concentration. The real nebulae would be
actually nothing more than groups of united single stars, but they are
situated possibly so far away in space that their resolution into single stars
would not be possible with the instruments available.

Herschel later modified the scheme quite a bit. It was improved according
to the impression of newer discoveries in one essential regard, and in so
doing he arrived at a wide-reaching interpretation concerning evolution: a
theory of stellar birth. On 13 November 1790 he found a highly peculiar
object, a star which was surrounded by a faint nebula. The star and nebula
were so closely bound to each other that Herschel assumed there was
necessarily a physical connection. From this came the idea of the existence
of a luminous material which is not made up of single stars and occurs

isolated from stars in space. Herschel mentions the Orion Nebula as an example of this self-luminous material. All considerations led him to the interesting conclusion that a star would result from the concentration of such a nebula, and the masses of finely mixed nebulosity made up the stuff from which the stars originate.

In 1811 Herschel finally put together all these indefatigably made observations, which were crowned by so many interesting discoveries. He made a classification scheme with thirty-two types of nebulae, supposedly representing an evolutionary sequence.

Evolutionary processes in the cosmos form part of the most complex questions of natural scientific research. Their scientific investigation requires deep understanding of the nature of the objects and a knowledge as complete as possible of all parameters which determine their qualities. On that score the astronomy of Herschel's time was not very advanced. As a result it is not surprising that Herschel failed to arrive at a permanent scientific understanding of stellar evolution, in spite of the installation of his large instruments and his untiring activity. Herschel did, however, with his wide-ranging investigations, help to diminish the importance of the metaphysical proofs in astronomy. He set up evolutionary thought as the scientific order of the day and thereby began a great new program of research which paved the way far into the future for astronomical research.*

While Herschel still continued with his investigations, the Frenchman P. S. Laplace published a cosmogony in his 1796 work *Exposition du système du monde* (Treatise on the System of the World), which consistently renounced the act of Divine creation and is to be compared in its general historical significance with the work of Kant. Laplace's theory had as its foundation the already existing central Sun, whose rotation was assumed to have caused rings of material to be thrown off from its atmosphere. These rings condensed to form the planets. In this work of Laplace the results of William Herschel were already being used and applied.

How did contemporary research regard all these bold ideas? As already emphasized, most of the scholars did not share Herschel's views. On this score one must not forget that the idea of an evolving cosmos was not only a scientific problem, precisely like the situation with regard to world view at the time of Copernicus. The introduction of evolutionary thought was a progressive step of great general significance. In the preface to his treatise on the natural history of the heavens Kant tried to make peace between the ruling ideology and his hypothesis and posited the evolution of the universe, regardless of whether his hypothesis proved the independence of nature

---

* It is sufficient to say that Herschel was not the first to demonstrate evolution in the cosmos. The ancient Greeks spoke of the 'incorruptibility of the heavens', and Renaissance scholars actively debated this issue. Tycho Brahe found from observations of the supernova 1572 that its lack of a measurable parallax placed it in the starry realms. Thus, two centuries before Herschel there was proof that the heavens change. (Tr. note)

from Divine Providence. For the most part eighteenth-century society had as yet no general interest in the assimilation of evolutionary thought. Where feudalism reigned it tried to cement its position of power with the conviction of the once-created, eternally unchangeable 'best of all worlds'. Should even the origin of the world be based now on natural forces and find itself in a permanent evolutionary process under the influence of these laws?

In this historical situation bold evolutionary concepts could not be irrefutably applied to cosmic bodies; evolutionary thought was introduced to astronomy only reluctantly and against the resistance of scientific and other forces. The German W. Olbers said what many astronomers firmly believed when he asked: 'Is this idea established or proven?' And he also gave an answer:

No, hardly. If we grant Herschel all his assumptions, what follows from his observations is nothing more than the existence of nebulous stars; there are stellar systems in which the stars are packed more closely to each other than in other systems. He has observed nothing further; everything else is only conjecture, and as I believe, somewhat unfounded, risky inference from his observations. Even change, which nature exhibits here on Earth on a small scale, will operate on the whole in the heavens, and may we also not wonder if all stellar systems are built according to one model?[12]

On the basis of possible errors of calculation Olbers rejected celestial mechanical arguments which spoke in favor of changes in the planetary system. With regard to evolutionary theory he thus took an outspoken, conservative stand. Olbers took the results of astronomical research already achieved to be sufficiently precise to justify final conclusions. Considering the authority which Olbers enjoyed in wide circles, the statements which he also repeated and expanded upon in popular treatises worked effectively against the assimilation of evolutionary thought.

In Göttingen, G. C. Lichtenberg, the important advocate of the German Enlightenment, believed differently. He advocated the evolutionary principle with energetic conviction, wrote a treatise *Ueber das Weltgebäude* (Concerning the Structure of the Universe), and had this opinion of Herschel's evolutionary concept: 'Although there remains a lot to be desired in these thoughts, it would indeed be difficult for someone who might not wonder about the immeasurable, because likelihood is enough for the assistance of speculation.'[13]

In the end, Herschel received little unrestrained acclaim by his contemporaries for his ideas on evolution in the universe. If someone admired him it was because of his untiring efforts, his numerous successes in traditional research areas, and even because one could not easily ignore the fascination associated with his results; he found hardly an imitator, and no spirited youths who would have enthusiastically followed in his tracks. Thus Herschel stood alone with his most genial works, his train of thought freely roaming far from the main areas of research. Science rarely attacks problems on command with all established intensity if a possibility of solution is not yet to be had. At the same time, contemporary astronomy intricately involved itself with other extensive undertakings. The validity of motion of heavenly bodies suggested itself as important. This research promised brilliant discoveries. Above all, however, a marked social need existed for discoveries in this area (see pp. 49 ff.). As a result this branch of astronomy enjoyed state funding. Thus celestial mechanics and positional astronomy became part of the official disciplinary sections of astronomy at the turn of the nineteenth century, while the questions of the structure and development of the universe remained for the time being only secondary concerns.

Marx and Engels attribute the greatest significance to the prevalence of evolutionary thought in the natural sciences. Engels especially occupied himself with the manifestation of evolution of heavenly bodies. He saw important elements for the dialectical-materialistic interpretation of nature in the suppositions and hypotheses which continually entrenched themselves in the course of research. The numerous related notes in his fragment *Dialektik der Natur* (Dialectics of Nature) are impressive proof of this, as these solely scientific points of view found their place in the Marxist philosophical theory of evolution.

THE MOTION OF CELESTIAL BODIES

The development of astronomy in the first half of the nineteenth century reached a practical and theoretical understanding of positional changes of objects in the heavens and the fixing of an astronomical coordinate system in the form of precise stellar positions. At this time the system of celestial mechanics founded by Newton was brought to perfection. The astronomy of this epoch owed its greatest successes to celestial mechanics and the ever-more-accurate stellar positional measurements, while conclusions concerning the physical constitution of the celestial bodies were hardly possible with the available observational instruments. The incontestable triumphs of celestial mechanics had another consequence too: they fostered the absolute importance of this branch of astronomy. This later became a very serious drawback for the subsequent development of astronomy.

The theoretical derivation of the laws which the motions of the heavenly bodies follow was beset by practical problems; time and again the practical needs forced theoretical refinements, and the increased accuracy of theory stimulated the development of still more accurate measurements. The German astronomer F. W. Bessel thus formulated a dialectical interplay between theory and measurement: 'One actually lacks the opportunity of building a theory of a phenomenon beyond the limits within which our perception of a phenomenon is confined; and there is just as seldom an occasion at hand to improve the perception, if the theory with which one can compare it is still in a very crude state.'[14]

Now the theory was resolved in a satisfactory way by means of the Keplerian laws of planetary motion. With the help of the *Rudolphine Tables* it was possible to calculate with sufficient accuracy the approximate elements of the planetary orbits and the future planetary positions. The masses of the planets could be regarded as very small compared to the Sun's mass, and the orbits – ellipses of very small eccentricity – were consequently considered to be adequately approximated by circles.

Theoretically, it was clear that the planets mutually influence each other and consequently that the motions experience certain perturbations. This problem, however, had for the time being only a small practical value; the consideration of more than two bodies played a role in astronomy solely in the treatment of satellite motion. The development of perturbation calculations and the investigation of three- and multi-body problems resulted in a period of intense theoretical studies. All great mathematicians after Newton concerned themselves very directly with these questions and created an important theoretical foundation upon which astronomy could later rely. The principal works of this epoch are J. L. Lagrange's *Mécanique analytique* (1788) and above all Laplace's *Mécanique céleste* (1799–1825). Laplace's work primarily was a second edition of the famous Newtonian treatise, enriched and refined with many new results. The work comprises five volumes and contains a detailed account of the state of mechanics, its historical development, and application to celestial bodies up to the date of publication.*

The complicated measurements concerning mutual perturbations of members of the solar system, which Lagrange and Laplace worked out, established an extraordinarily important result: they proved the stability of the solar system. Previously, one might have believed in general that the mutual perturbations of the planets – as small as they might be – must eventually result in irreversible consequences for the whole system, but Lagrange and Laplace showed that the effects remain within certain limits and the whole system is stable.

The strictly analytical solution of the three-body problem by the French mathematician J. L. Lagrange is one of the most interesting results of the theory of motion of celestial bodies for one special case. Lagrange showed that a strict solution of the problem exists if one of the three bodies is found in a very specific geometrical orientation with respect to the other two – if it moves along on one of the five libration points. When this work was published in 1772 it brought great renown to its author, but it appeared to

---

* One interesting story concerning the *Mécanique céleste* involves Napoleon, who said to its author: 'M. Laplace, they tell me you have written this large book on the system of the universe, and have never even mentioned its Creator.' Laplace retorted, 'I have no need of that hypothesis.' (Tr. note)

have no practical importance. It was later found that within the planetary system there are celestial bodies that very closely satisfy the just-mentioned condition, the group of Trojan asteroids. The first of these was discovered by Max Wolf in 1906.*

That only theoreticians concerned themselves with the problems of perturbation calculations had an undesired result for practical astronomy. The mathematicians developed the methods of orbit calculation of celestial bodies in great generality, but in their works they hardly considered the question of practical calculation – very important for the application. For example, Lagrange derived formulas which in their general form allowed the determination of an elliptical orbit on the basis of a few observations, but they were so complicated and unfit for use that they were hardly used, and again only theoreticians made note of these drawbacks.

The discovery of the planet Uranus in 1781 gave no opportunity to work out new and more easily used methods of planetary orbit determinations. The extraordinarily slow motion of this planet, and the small eccentricity and inclination of the orbit permitted the approximate orbit determination with sufficient accuracy according to the old procedures.

A similar situation was found in the area of calculating cometary orbits. The problem was indeed more difficult; the exact form of the orbit was not known, nor were the inclinations, nor some of the periods of revolution. The relatively small number of comets then known did not allow any final conclusions to be made concerning the membership of the comets to the solar system. Progress in this area was made by a contemporary and friend of Isaac Newton. The English astronomer E. Halley compiled a great amount of recorded observations of comets for the calculation of the orbits, which he gleaned in part from a number of old sources very difficult to find. He used the Newtonian Law of Gravity for the calculations. As a result he made the astonishing discovery that several comets exhibited very similar orbits. He hypothesized that it was one and the same comet and concluded this on the basis of the noteworthy cometary appearances of 1456, 1531, 1607, and 1682.[15] Halley went further and predicted the reappearance of this comet in 1758. The German farmer and astronomer J. G. Palitzsch actually discovered the comet on Christmas Day 1758 and confirmed the correctness of the interpretation of Halley, who had died in the meantime. It was thereby proven that Halley's Comet is a member of our solar system; but one could not propose a generalization of this type of result. Later almost all great mathematicians dedicated themselves closely along the lines of the theory of cometary orbits, among them such prominent theoreticians as A.-C. Clairaut, L. Euler, and P. S. Laplace.

The disadvantage of the different methods for the determination of cometary orbits was that an extraordinary ability for calculation was

---

* There are about 700 Trojans in the Achilles group (60° ahead of Jupiter) and about half that number in the Patroclus group (60° behind Jupiter). (Tr. note)

required – the methods were complicated and slow. The more frequent discoveries of comets became a serious impediment to research for the small number of astronomers capable of calculating orbits with the practical procedures. Like the others who concerned themselves with comets, W. Olbers also experienced this problem. In his calculations he succeeded in deriving an important simplification of the procedure. Instead of calculating the orbit on the basis of the Keplerian Law of Areas, Olbers represented the resultant portions of the sections of the cometary orbit and the radius vectors as triangles, which were made up of the radius vectors and the chords between two positions of the orbit. The chord sections are required to match the corresponding time intervals. In the same way Olbers took care of the motion of the Earth in its orbit. From this he could derive relatively simple formulas, allowing a convenient and quick method of calculating the cometary orbit. The results of Olber's methods were presented in 1797 to the Göttingen Gesellschaft der Wissenschaften (Scientific Society) in his German-language 'Treatise concerning the easiest and most convenient method of determining the orbit of a comet'. The treatise is a classic of celestial mechanics and was published in 1797 by F. X. von Zach in Gotha. The small but significant work went through many editions and was also translated into other languages. It united the results of celestial mechanics with the promotion of practical astronomy; it can be used in an applied form and thereby it rationalized the work of astronomers.

Due to the rapid development of telescopes 351 comets were discovered during the nineteenth century alone, nearly twice as many as in the previous three centuries. The growing abundance of material and the continuously improving theoretical grasp of the celestial mechanical behavior of comets gave birth to good suppositions concerning the understanding of statistical regularities of the orbit forms.

Particularly noteworthy was the determination that the eccentricities of almost all of the comet orbits are less than 1; the few orbits with present eccentricities greater than 1 have not always been hyperbolic – in all cases they used to be parabolic.* Toward the end of the nineteenth century the original orbital elements of comets with hyperbolic or parabolic orbits were calculated with the application of perturbation determinations. A result from these calculations was that the original orbits turned out to be near-parabolic ellipses. From this an important conclusion could be made: all comets are members of the solar system.

In 1818 the German astronomer J. F. Encke succeeded in calculating the orbit of a comet previously observed many times. This object, now known as Encke's Comet, was shown to have an astonishingly short period. With

---

* Elliptical orbits have eccentricities greater than or equal to o, but less than 1; parabolas have eccentricity exactly 1; hyperbolas have eccentricities greater than 1. Comets with elliptical orbits are necessarily bound to our solar system; those with hyperbolic orbits necessarily escape. (Tr. note)

an orbital period of 3.3 years it remains the comet with the shortest known orbital period around the Sun. In connection with Encke's discovery several more short-period comets were found. In a great many of the cases the determination of the orbital elements showed that many of the semi-major axes correspond to the orbital radius of the planet Jupiter. In this way the Jovian family of comets was discovered by statistical means.

The increased knowledge concerning comets also systematically helped eliminate superstitions concerning them. One of the important results which helped to undermine the fear of comets and also increased scientific knowledge was associated with the observation of the short-period comet known as Biela's Comet. The comet was expected to appear in the sky in 1846, on account of its known orbital period of 6 years 9 months. However, it exhibited a bifurcation, which had never previously been observed. During its next approach in 1852 both portions were considerably further apart. By 1866 the comet could not be found at all. When G. V. Schiaparelli proved in the same year that the orbit of the Perseid meteor shower was identical to that of the periodic comet Tuttle, astronomers found themselves on the right track to discovering the clue to the disappearance of Biela's Comet. H. L. d'Arrest, who had discovered the splitting of Biela's Comet in 1846, and E. Weiss showed that a meteor shower would stream forth from the direction of the constellation Andromeda, having the orbit of Biela's Comet. They predicted the appearance of an active meteor shower for 28 November 1872, the date of the next passage of the Earth through the orbit of the former comet. The prognosis bore itself out with remarkable precision: in Italy four observers counted 33 400 Andromedid meteors in $6\frac{1}{2}$ hours. It was thereby clearly proven that the comets, which were associated with disastrous fate for the Earth's inhabitants, were heavenly bodies subject to the laws of nature. The relationship between the meteor showers and the comets was also irrefutably proven for a few cases.

The growing number of comet discoveries had led to the practical solution of the problem of determining an orbit; it similarly led to practical demands and also eventually to new discoveries concerning the solution of this problem for the planets. The minor planets played a corresponding role in these discoveries.

THE MINOR PLANETS

The discovery of Uranus had once again made astronomers particularly aware of a known numerical pattern. This pattern had been noted by Kepler, and also by C. Wolff in 1724. It was then expressed in mathematical form by J. D. Titius and J. E. Bode. The distances of the planets from the Sun follow a distance law, which has come to be known as the Titius–Bode Law. According to the law one relates the distances a of the respective planets from the Sun* according to the equation $a = 0.4 + 0.3 \times 2^n$, if one takes for Mercury, Venus, Earth, Mars, Jupiter, and Saturn the exponents $n = -\infty, 0, 1, 2, 4,$ and 5. It had long been noted that a gap existed between the planets Mars and Jupiter, i.e., that no planet was known whose distance corresponds to the value $n = 3$ in the Titius–Bode Law. After the distance corresponding to $n = 6$ matched that of Uranus, many felt that another planet was to be found between Mars and Jupiter. The astronomer Zach so strongly believed in the existence of this planet that he estimated the distance of the body and other data by means of the equation and the Keplerian laws and in 1785 deposited these data in sealed envelopes with colleagues in Berlin, Dresden, and London.

Because all planets approximately move about the Sun in a plane, the ecliptic, this was also expected for yet undiscovered planets. As a result very accurate star maps of the zodiac became necessary in order quickly to discover the comparatively faint planets. The expectation of a planetary find made the toilsome celestial surveys attractive to many astronomers. In Lilienthal (near Bremen) German astronomers met at the observatory of J. H. Schroeter in the autumn of 1800 and founded the Vereinigte Astronomische Gesellschaft (United Astronomical Society), with the explicit purpose of collaborating with foreign astronomers on the production of zodiacal star charts. Many astronomers proposed there as co-workers for the undertaking had already accomplished important preparatory work on their own initiative, especially the Italian astronomer G. Piazzi, who had concerned himself with a 'revision' of the sky starting in the 1790s. As a result of his systematic observations Piazzi found an object on New Year's Day of 1801, the very beginning of the nineteenth century, which could not be a fixed star on the basis of his reckoning, and which he subsequently concluded to be a comet. On the next evening Piazzi had the opportunity of measuring the subsequent motion of the object against the background of stars. He constantly monitored the new object until 11 February, by which time he confirmed his supposition that it was a planet. Then, because of sickness, it was not possible for Piazzi to continue observations. When the news of the discovery reached northern Europe a period of bad weather hindered further monitoring of the object; and finally when there was another opportunity to make observations the new object could not be found. The relatively faint luminosity of 9th magnitude gave the little star secure shelter from the eyes of astronomers, especially when it was in a densely stellar region and the available star charts could not aid the recovery of such faint objects.

The surest means of recovering the missing heavenly body was an orbit determination.

---

* Here the distance from the Earth to the Sun – the astronomical unit – is used as the unit distance. (Tr. note)

Under the assumption of a circular orbit F. X. von Zach carried out the calculation of the ephemeris for the months of November and December 1801 in the solid conviction that the new heavenly body was a planet. However, at the same time the young German mathematician C. F. Gauss had learned of the problem of an ephemeris calculation based on the comparatively few geocentric positions of the object by means of the rapid distribution of the reports in the newly founded journal *Monatliche Correspondenz zur Beförderung der Erd- und Himmels-Kunde* (Monthly Correspondence for the Advancement of Geography and Astronomy). He promptly addressed himself to this task. Without assuming the position of the body in its orbit Gauss succeeded in calculating an ephemeris on the basis of the small arc which was known by observations. This led to its rediscovery by Olbers in Bremen on 1 January 1802, precisely a year after the first discovery of the object (subsequently known as Ceres). The right ascension of the Gaussian ephemeris proved to be 7° different than the value of Zach. The astronomers were exceedingly impressed by this achievement, for the ellipse calculated by Gauss represented the observations of Piazzi astonishingly well. For the time being Gauss did not communicate details of his method, but everyone was quite convinced that a new method had been developed.

Gauss had actually attempted a general solution of the problem of deriving good ephemerides from a few observations, without having to resort to random assumptions. He had completely finished this task by the end of 1801, as a manuscript sent to Olbers proves. The discoveries of subsequent minor planets during the years 1802, 1804, and 1807 were a welcome opportunity for Gauss to permanently improve his method. When Olbers discovered the asteroid Vesta in 1807 and sent Gauss the observations, it required only a few hours for Gauss to calculate the orbital elements. The secret of his method was soon to be disclosed: in 1809 Gauss published his procedure in the classical work *Theoria Motus Corporum Coelestium in Sectionibus Conicis Solem Ambientum* (Theory of the Motion of Celestial Bodies which Revolve about the Sun in Conic Sections). Also, the well-known 'method of least-squares' is used in this work for the first time, having been developed by Gauss in 1794. With the application of this method he succeeded in using all the observations for the calculation of the orbit, and not only specifically selected data.

Like Olber's method of calculating a comet orbit, Gauss' theory was in great contrast to the theoretical works of the successors of Newton, above all J. L. Lagrange, P. S. Laplace, and A.-C. Clairaut; Gauss' theory addressed the practical problem of carrying out the measurements. In this respect it also represented an important simplification in the work of calculating astronomers.

On one occasion Gauss was able to determine an orbit in one hour by means of his *Theoria Motus*; the famous Leonhard Euler required three days of strenuous work for the same problem as solved by his own methods. The

Fig 5. C. F. Gauss

rationalization of the work through the creation of practical calculating procedures was of greatest importance for the rapid development of astronomy and the close interaction between observation and theory. Already in 1802 Zach wrote concerning this progress that months and days were required where several decades or years were previously necessary. Toward the end of the eighteenth century there were only four or five astronomers in all of Europe who could calculate the perturbation equations for the asteroid Ceres in several months, whereas 'these days there are perhaps more than a dozen young and clever men who could complete such work in a few days'.[16]

The development of the Gaussian *Theoria Motus* is an especially graphic example of the stimulation of theory by practical needs and of the uses which practice draws from theory. In spite of all subsequent refinements the method derived by Gauss has remained the most widely used procedure for orbit calculation. Several improvements, which above all led to further refinements of calculation, were undertaken later by Gauss' student J. F. Encke. Subsequent variations of the Gaussian method were derived by T. von Oppolzer, F. Tietjen, P. Harzer and others.

Incidentally, the discovery of Ceres led to a curious and little-known controversy between the astronomers and the young G. W. F. Hegel. Hegel had put forth the formal proof in his dissertation, which had just been published, that there can only be seven planets in the solar system.[17] Naturally, the discovery of Ceres was an embarrassing blow for the philosopher, and the astronomers wasted no time in making the most of it. Zach called Hegel's dissertation 'literary vandalism' by people 'who should first learn before they can teach'. The later mistrust of numerous scientists for philosophy altogether, which was greatly regretted by Marx and Engels, did not have its ultimate roots in the natural-philosophical refuse of Hegelian philosophy; Zach uncategorically states this when he opposes the consideration of Newtonian physics as an Hegelian 'hyperphysics' and emphasizes that Newtonian physics 'always gives occasion to the shining discoveries in the universe', while Hegel's philosophy 'not only fails to produce the most trifling discovery, but actually prevents its discovery'.[18] On the other hand there were Hegel's later great achievements, upon which Marx and Engels could develop the materialistic dialectic, and they are therefore to be counted among the most important theoretical sources of Marxism. It was Hegel who first described the general dialectical forms of motion in a comprehensive and well-known way.

The great role played by the asteroids in the rapid development of theory and observation becomes clear if one keeps in mind the history of subsequent discoveries of asteroids.

On 28 March 1802, shortly after the rediscovery of Ceres, Olbers surveyed the region of the constellation Virgo, where he had seen Ceres in January. He noticed a star which had not been situated there at the time of the Ceres discovery. Over the course of only a few hours Olbers was able to detect a motion of this object against the background of fixed stars. After several days it became clear that it must be a planet. Within the shortest possible time Gauss determined the elliptical orbit of this new planet (Pallas) according to the procedure developed by him, and his ephemeridal predictions agreed very well with the first subsequent observations. Later the orbital elements were greatly improved; above all the relatively strong perturbations by Jupiter's mass had to be taken into account.

The most interesting result of both the orbit determinations for Ceres and Pallas was the fact that they had essentially the same period of revolution about the Sun, and their so-called orbital nodes were very close to each other. For Ceres one found an orbital period of 1681.4 days and for Pallas 1686.3 days. In addition to this other very interesting results followed from the observations of J. H. Schroeter and William Herschel; both concerned themselves with the question of the dimensions of the new heavenly bodies. Their results hardly agree, but it was agreed that the new heavenly bodies had small masses in comparison to the other known planets of the solar system. With the noteworthy orbital orientation and these quantitative determinations Olbers set forth his hypothesis that these bodies must be leftover pieces of some originally larger planet. From that he further concluded that numerous other pieces of the original large planet were certain to be found in the future. This prognosis was confirmed when the German astronomer C. L. Harding found a third minor planet, Juno. Olbers himself enlarged the series of these discoveries with the identification of Vesta in 1807 as yet another object. The largest of the minor planets were thereby tracked down, and for a long time there were no other finds, although Olbers and other astronomers were firmly convinced that many more asteroids were concealed in the swarm of fixed stars.

Not until 1845 did a new epoch of asteroid discoveries begin; it was issued in by the German amateur astronomer K. L. Hencke, who discovered the minor planet Astraea, and in 1847 found Hebe. In the next twenty years a comparatively small number of astronomers found more than eighty minor planets by means of systematic telescopic observations. At the beginning of the 1890s more than 300 planetoids were known. However, a flood of discoveries was at hand when Max Wolf undertook detailed photographic investigations as a means of discovering minor planets. At the same time Wolf pursued the goal of rediscovering 'lost' asteroids, especially when these missing objects had not been subjected to orbit determination. Wolf photographed all regions of the ecliptic in which the already known minor planets attained positions of opposition from the Sun, for there the brightest luminosities were to be measured. He guided the instrument according to the apparent daily motion of the fixed stars, so that a minor planet would necessarily leave a streak upon the plate. When Wolf undertook the first photographic investigation toward the end of 1891, he succeeded in one stroke (with an exposure time of two hours) in rediscovering the 275th minor planet (Sapientia) and he also discovered a previously unknown asteroid. In the following year in Heidelberg Wolf found seventeen new minor planets by himself.* The era of photographic discovery of minor planets was thereby opened. As a result the number of known asteroids soon topped a thousand.

The astronomers had become burdened with an extraordinarily large observational and calculatory task because of the minor planets. They nevertheless carried out these tasks, for the minor planets claim a particular interest in various respects. On the one hand, because of their small masses, the calculation of their orbits is a very important problem in celestial mechanics, for they can be used to a certain extent as celestial mechanical controls; on the other hand they became of interest to an increasing degree

---

* Wolf discovered well over a hundred asteroids. We should also mention a few other accomplished asteroid hunters. Palisa at Vienna discovered 120 of his own, but after a period of rivalry with Wolf they collaborated; Wolf was given the task of discovering the asteroids at opposition, and Palisa followed them out as far as he could with the large 27-inch refractor at Vienna. Other asteroid hunters of note include A. H. P. Charlois of Nice and C. H. F. Peters of Clinton, New York. (Tr. note)

for the cosmogony of the planetary system, something Olbers had already suggested. Finally, a series of other important astronomical quantities were measured with the help of the minor planets, like, for example, the solar parallax, by means of the minor planet Eros, discovered in 1898 (see p. 41). Also, precise mass determinations of the large planets were achieved with the help of precise asteroid observations.

STAR CHARTS AND STAR CATALOGS

A knowledge of stellar positions was an incontestable prerequisite for the precise determination of planetary, cometary, and asteroidal orbits. Moreover, in 1718 E. Halley discovered changes of stellar position, the so-called proper motions, which likewise required the careful measurement of stellar positions. As a result the production of star charts and star catalogs became a primary concern of astronomers in the eighteenth and nineteenth centuries.

This involved compiling the most accurate stellar positions possible for the greatest possible number of objects. Because both types of data could not be simultaneously realized, two types of compilation developed, (1) the general survey, in which the accuracy must be directed toward the identification of the objects, and (2) the catalog with relatively few objects, which have, however, such accurate positions that they are used as a foundation for the derivation of other stellar positions and represent the astronomical coordinate system.

The oldest positional catalog which is also of value to modern research was put together by the first director of Greenwich Observatory, J. Flamsteed. Numerous other observers followed in his footsteps. Among them Piazzi, Lalande, and Bessel struggled in vain as individual researchers to overcome the sheer volume of material. Zach had already realized that a careful documentation of stellar positions down to the faintest possible stars could only be realized within a reasonably projected time span by a collective undertaking. His plan for a zodiacal survey with the collaboration of astronomers around the world (which had been discussed at Schroeter's observatory in 1800) was regrettably not realized. However, the project was not forgotten. F. W. Bessel, who had begun his practical astronomical career with Schroeter in Lilienthal, resumed the plan once again, though not until a quarter century later. In a detailed letter to the Royal Academy of Sciences in Berlin in October 1824 Bessel underscored the need for a comprehensive stellar catalog, as follows:

If a planet or comet is to be observed away from the meridian, this endeavor will meet with greater success if there are always many well determined stars in the vicinity: if one seriously wants to aim at discovering all the principal planets that belong to the solar system, the complete cataloging of the stars must be carried out. ... The hope of discovering several new planets in this way must be considered very well established.[19]

Fig. 6. F. W. Bessel

When Bessel wrote this letter he had more than a theoretical conception of the project. Three years of dedicated observations already lay behind him; he had measured with a meridian circle all stars in a 30° wide zone centered on the celestial equator down to a luminosity of 9th magnitude. From this he drew up a map of 1 hour of right ascension in length, which he submitted to the Academy as a model. Thus the undertaking of the 'Academic Star Charts' was begun. The complete work was not published until 1859. Bessel continued his zonal observations until 1835, and from about 75 000 individual observations he derived the positions of 31 895 stars. Incidentally, as a result of these observations he discovered in 1821 a 'measure of error' as it were, the 'personal equation' as he called it – his assistant measured the contact of the stars with the threads in the field of view of the eyepiece consistently one second later than he did. The knowledge of this time difference is understandably very important in evaluating the results. In 1889 J. A. Repsold largely succeeded in eliminating the 'personal equation' through the construction of an 'impersonal micrometer'.

Besides Bessel, numerous other astronomers concerned themselves with the production of extensive star catalogs. We must at least mention here the copious works of S. Groombridge, G. B. Airy, R. C. Carrington, and J. von Lamont. The last-mentioned observed stars down to 10th magnitude for the zones chosen by him and used 80 000 individual observations for his catalog.

However, the greatest undertaking, fundamental for all subsequent works, was the comprehensive stellar survey carried out by the German astronomers F. W. Argelander and E. Schönfeld at the Bonn Observatory, the so-called *Bonner Durchmusterung* (*BD*), which also lists the estimated magnitudes of the stars along with the approximate positions (see pp. 88 ff.). The northern *BD* alone contains altogether the approximate positions of 324 198 stars, to which Schönfeld added 133 659 stars down to declination −23°. The rest of the southern sky is contained in the *Cordoba Durchmusterung* (613 953 stars).

The zonal undertaking of the Astronomische Gesellschaft is a still more extensive project and was only realized through the collaboration of many observatories. The efforts of all participants went on for decades. This gigantic project was proposed by Argelander, the initiator of the *BD*. All stars between −2° and +80° declination down to 9th magnitude were to be observed. As a student of Bessel who upheld the Besselian tradition, Argelander honored his mentor, but in Germany the growing tradition of positional astronomy had reached maturity; it had indeed become an overgrown part of the astronomer's life and only grudgingly allowed room for new ideas. Altogether seventeen observatories worked on the zonal undertaking of the Astronomische Gesellschaft, among them the observatories of Pulkovo, Helsingfors, Bonn, and Berlin.

The zonal undertaking was supposed to create a uniform and valid foundation for all future observations. Therefore, it was planned to repeat the whole programme approximately half a century later; stellar proper motions were to be derived from the data.

The result of the first survey was set down in a catalog of fifteen volumes (*AGK 1*), which contained data on a total of 150 000 stars. An extension of the catalog, which covered the declination range −2° to −23° and which appeared in 1887, contained data on more than 50 000 stars. The second phase of the task was begun in the 1920s. This time the advantages of photography could be employed.

Fig. 7. F. W. Argelander

Back in 1887 an International Congress for Astrophotography was held in Paris to organize the production of a mapping of the whole sky by means of photography. Extensive preliminary investigations were carried out for this purpose, with the participants agreeing upon uniform instrumental equipment for all collaborators, uniformly suitable plate materials and formats, and obligatory exposure times. One hundred and fifty catalog volumes were to contain the coordinates of several million stars. For this purpose 22 000 charts were planned, with stars down to 20th magnitude. Historically, however, the undertaking was prematurely attempted for knowledge concerning the application of photography was very limited even in the 1920s. Not until 1970 was the photographic star mapping finished.

By 1925, however, valuable experience had already been gained which could be used for the second phase of the *AGK 1*. This undertaking planned to limit the whole northern sky to 2000 plates. In addition to this, 13 747 standard stars were chosen whose positions were to be determined in the classical fashion with a meridian circle, in order to transform the relative positions on the plates to spherical coordinates. The 13 747 standard stars had to be fixed as closely as possible to a system of fundamental stars whose average positions were known with the greatest possible precision for a given epoch.

The fundamental catalogs have their own history. The most important work of this kind in the first half of the nineteenth century was the *Fundamenta Astronomiae* of F. W. Bessel. Bessel worked seven years on this classical work, which was published in 1818. It contains the average positions of 3220 stars, is mainly derived from the precise observations of the great English observer, J. Bradley, and is reduced to the year 1755 (epoch 1755). Due to the prevailing conditions in Germany, this important work was only published with the greatest difficulties and through the authoritative initiative of B. A. von Lindenau. It includes all formulas and tables required for the reduction of other observations to the fundamental system. With a consideration of new findings, A. von Auwers once again adopted Bradley's observations; the result was the *New Fundamental Catalog of the Berlin Astronomical Yearbook*. Continual improvements to the system of fundamental stars finally led to the presently used fourth fundamental catalog (*FK 4*).

The adoption of a system of fundamental stars is an extraordinarily complicated task, for it includes permanent research on a whole series of special areas of activity. In order to derive the average position of a star for an epoch with the highest possible precision, the values read right off the telescope must be 'reduced' by extensive mathematical operations. The value read off the telescope by no means represents the best possible position of the star; it includes numerous biasing effects. As a result a controversy arose as to which reduction elements should be considered and in which way they should be applied. For example, one needs to consider the refraction in the atmosphere and the periodic shifts of stellar positions due to aberration, as well as the change of the origin of the coordinate system (precession). However, these effects were only gradually properly considered. Refraction depends on the temperature and air pressure in a complicated way. The theory of refraction indeed begins in antiquity with Ptolemy, but the elementary treatment of a scientific theory of refraction was not set down until the discovery of the law of refraction (W. Snell, 1602). Then Newton gave us the basis for a modern refraction theory. After Newton a number of people worked on this important subject, among them D. Bernoulli, T. Mayer, J. H. Lambert, L. Euler, J. L. Lagrange, P. S. Laplace, and F. W. Bessel. The research necessary for other reduction elements was similarly complicated. It was finally realized that no instrument could be as perfect in practice as was hoped. As a result the problem arose of determining the instrumental errors by practical tests, and of considering the reduction of the observations in a suitable way. 'Every instrument is made twice over in this way', wrote F. W. Bessel, 'once in the shop of the master from brass and steel, a second time, however, by the astronomer on paper, through the application of necessary corrections, which he obtains in the course of his investigations.'[20] Important progress was made on the improvement of the testing and theory of astronomical instruments by Bessel himself.

In astronomy, in contrast to most of the natural sciences, older measurements can remain of value for the investigation of certain phenomena. This is particularly true for positional data. Because it is very time consuming to locate required data for a specific purpose from the numerous, highly inaccessible older catalogs, and to take into account the respective epochs, A. von Auwers first expressed the intention in 1878 of putting all works still of value for actual research in the area of positional measurement together into one work. In 1897 this idea took concrete shape when F. Ristenpart proposed 'that references should be set down to all star catalogs, in which any stars are found, in some kind of collection'.[21] In 1900 A. von Auwers communicated this plan of a 'catalog of catalogs' to the Berlin Academy of Sciences, and a 'history of the starry sky' was formally proposed. A gigantic program lay before the Academy. All stellar positions measured with the aid of meridian instruments from 1750 to 1900 were to be reduced to epoch 1875 and converted to a uniform system. Approximately one million individual measurements from about 450 star catalogs were used for this purpose. The undertaking was considerably hindered by the First and Second World Wars, and the last volume was not published until 1966.

Celestial mechanics and positional astronomy led to brilliant discoveries in the nineteenth century: the planet Neptune, the determination of the astronomical unit, the distances of the stars, the investigation of double stars, and, last but not least, important applications in the areas of geography and geodesy. These are some of the triumphs with which the untiring endeavors of astronomers were rewarded.

THE DISCOVERY OF THE PLANET NEPTUNE

In spite of the fact that the approximate orbital elements of Uranus were easy to calculate on the basis of the Keplerian laws, the attention of many astronomers was directed toward pursuing the details of its motion. Here the astronomers were in luck, because several positions of Uranus were at their disposal which dated from the time before its discovery. The oldest of them came from the year 1690, at which time J. Flamsteed had registered the planet as a fixed star without suspecting that it really was a planet. When Bessel worked through J. Bradley's observations for his *Fundamenta Astronomiae* he discovered that Uranus was measured no less than seventeen times between 1742 and 1762 for that project alone. The German astronomer Tobias Mayer had seen Uranus in 1756 and designated it as No. 964 in his list of zodiacal stars. Finally the Frenchman Lemonnier had also taken the planet to be a fixed star in December 1768.

Certainly, everyone did not agree that the old positions in every case were associated with this planet; the identification of the planet according to the backward reckoning of its movement from the approximate data could not be done with absolute certainty. When the Paris Academy offered a prize for a theory of Uranus' orbit, J.-B. J. Delambre, among others, looked for a solution. He was understandably very interested in older observations and was able to procure some of this information from G. C. Lichtenberg, who had access to the original observations of Mayer. Because of the results of his calculations, Delambre expressed doubt concerning the hypothesis that Mayer actually saw Uranus. The tables of Uranus' motion which Delambre worked out deviated up to 20 arc seconds from the observations up until 1811. In 1821 the French theoretician A. Bouvard made a new attempt. The observational material in the meantime had grown considerably. However, Bouvard's study also failed to reconcile the assumed or actual pre-discovery observations of Uranus. This astronomer then stated that all the observations could not be reconciled with a single system of orbital elements; he had to leave it undecided whether this was due to the inaccuracy of the older positions or if it really was caused by something else. Later Bouvard also guessed that the newly discovered planet was perturbed in its motion by a still more distant, undiscovered planet. Other astronomers also entertained such ideas; for example, F. W. Bessel expressed this opinion in 1823 in a letter to Olbers. Bessel had even begun to subject the available observations to a serious test, with the sole purpose of achieving 'the most beautiful enrichment of science'.[22] Altogether he could enumerate 361 observations since the time of the discovery of the planet, of which 143 had been made at the observatory directed by him in Königsberg. It was a difficult calculational task: 'The vein of pure gold . . . lies deep; we must continuously and furiously dig in the main shaft for it . . . because at this time the worker deep below sweats for only rock and poor earth. The smelting of this is a meager reward before the finding of the vein.'[23] Bessel and his student Fleming, who carried out the calculations, did not finish the mammoth program. Two other young astronomers, U. J. J. Leverrier in France and J. C. Adams in England, had also begun to work on the problem. From the beginning they advocated the thesis that the deviation of Uranus' actual motion from the theoretical ideas of its motion must result from perturbations from a yet-unknown planet, and they set about to derive the position of the unknown object from the measured perturbations. Adams finished his calculations in the autumn of 1845.*

From his results it was clear that the unknown planet could be found in the sky; he sent his results to J. Challis, the Cambridge professor of mathematics and astronomy. However, Challis dedicated himself to the task at hand with much less enthusiasm than the Astronomer Royal, G. B. Airy, who was also informed. Challis did not observe the region of the sky in question until late the following summer and actually saw the new planet on 4 and 12 August, as it turned out soon afterwards, but he had failed on these occasions to reduce the observations right away. A direct proof of the discovery of the planet was not possible through a star comparison, for suitable star charts were lacking. In the meantime U. J. J. Leverrier had also finished his calculations. Shortly before that J. G. Galle, who, along with J. F. Encke, was then employed at the Berlin Observatory, had sent a copy of his dissertation to Leverrier, among others. Leverrier passed along with his thanks his calculations of the position of the new planet on 1 January 1847, and asked that Galle 'might dedicate himself for a few moments to the search of a region of the sky where a planet could be discovered'.[24] The letter, dated 18 September, arrived in Berlin on the morning of the 23rd, so Galle made observations that very evening of the region of the sky in question. At the suggestion of d'Arrest, a student also present, Galle brought along some star charts belonging to the collection being published by the Berlin Academy that were available nowhere else. After several comparison views with the help of the appropriate chart, Galle discovered a star of 8th magnitude in the field of view of the 9-inch Fraunhofer refractor which was absent from the chart. In the presence of Encke the observation of this object was carried out throughout the night, without a trace of motion against the background of fixed stars being evident. However, the renewed observations the following night made them certain; the little star absent from the star chart had moved on, and was by all appearance the new planet of the solar system predicted by Leverrier.

Because the position of the newly found object differed by 55 arc minutes from the calculated position, several astronomers tried to attribute the whole discovery to pure chance. The mathematician C. G. J. Jacobi wrote quite justly in his notebook: 'One must wonder how such precise results could have been drawn from such sparse and uncertain quantities; this can only be attributed to the circumspect treatment of these data and to the exemplary application of all aids. Those who attribute the discovery to chance because the agreement is not closer than the nature of the thing

---

* Sir John Herschel, in an address to the British Association on 10 September 1845 said: 'We see it [a probable new planet] as Columbus saw America from the shores of Spain. Its movements have been felt, trembling along the far-reaching line of our analysis, with a certainty hardly inferior to that of ocular demonstration.' (Tr. note)

permits should indeed give permission for other chance discoveries to be made.'[25]

The positional calculation and actual discovery of Neptune was a triumph without equal for celestial mechanics. Although at the time of this scientific achievement no one doubted the validity of the Newtonian Law of Gravity, the understanding of the celestial laws had indeed been demonstrated in a particularly complete and convincing way as a result of this achievement. Astronomical prognoses were at that time nothing new for the researcher, but the discovery of a new large planet by means of the application of a theory understandably made a great impression in the widest circles and was of particular propaganda value, since the confirmation of theory through practice demonstrated the level of understanding of celestial mechanical laws.

THE ASTRONOMICAL UNIT

One of the important values for all of astronomy is the distance of the Earth from the Sun, which is represented by the measurable solar parallax.* Since the *relative* dimensions of the planetary system can be derived from Kepler's Third Law and the measurable periods of revolution of the planets about the Sun, the *absolute* orbital size of one planet is required in order to also determine the absolute orbit sizes of the other planets. The distance of the Earth from the Sun, the 'astronomical unit' (AU) is also necessarily the 'astronomers' yardstick' for the determination of all other distances in the cosmos; the distances of stars result from different relative measurements, so that the accuracy of all other distance methods are directly related to the accuracy with which the solar parallax is known.

However, a direct, accurate measurement of the solar parallax is not possible for various reasons. Therefore, attempts are made to measure the astronomical unit in a roundabout way with daily parallaxes of planets, especially when these become large angles, i.e., when the planets reach opposition. In addition, an accurate value of the Earth's diameter is required as a baseline for the trigonometric parallax measurement. The first reliable determination of the Earth's radius was made by J. Picard in 1669–1670; a scientific foundation for the measurement of the celestial regions was achieved simultaneously with this terrestrial measurement. The solar parallax was determined for the first time in modern history by the Frenchman Lacaille with the close approach to the Earth of the planet Mars in 1751. The result was approximately correct.[26]

The transits of the planet Venus across the disk of the Sun in the eighteenth and nineteenth centuries played a noteworthy role in the

---

* The solar parallax is the angle subtended by the Earth's radius at a distance of one astronomical unit. It is then half the angular difference of the Sun's positions as measured at both ends of a baseline the size of the Earth's diameter, at one Earth–Sun distance. (Tr. note)

determination of the astronomical unit. E. Halley proposed the observation of Venus transits from places of differing geographic latitude and the measurement of the parallactic shift of Venus in front of the solar disk; the solar parallax can be derived very precisely from this method, especially because the distance of Venus from the Earth during the inferior conjunction of the planet is reduced to about 45 million km. Venus transits are quite rare; in the eighteenth and nineteenth centuries a total of only four Venus transits took place, which were the object of careful measurement.

The transits of 1761 and 1769 were observed with considerable expense; numerous observatories outfitted expeditions in order to pursue the event from suitable places. The Russian and Swedish astronomers traveled around in their own countries to widely separated locations. The English and French astronomers undertook even more involved expeditions. One of the English expeditions set out to Bengkulu in the East Indies. But the English and French were at war with each other, and the English warship, on which the astronomers C. Mason and J. Dixon had hoped to safely reach India, was attacked by the French, and it had to turn back with the dead and wounded on board. The two astronomers finally did reach the Cape of Good Hope and carry out the measurements from this ill-equipped place, but only with difficulty. This greatly enhanced English prestige.

The French observations were organized by J.-N. Delisle, who had been in personal contact with Newton and Halley. The French travels also took shape with difficulty, owing to the political conditions. Pondicherry, a French settlement in India, destination of the astronomer Le Gentil, was still occupied by the English before his arrival, and he could only observe the Venus transit on the high seas during the return journey.

Both the English and the French astronomers were partly hindered in the observations by the effects of weather. Practically no experience was gained in observing transits. Above all, the appearance of the 'black drop' rendered the exact temporal fixing of the first contact more difficult. All in all the results were so scanty that everyone anxiously waited eight years for the next Venus transit.

The international research expeditions were much more extensive in 1769. Moreover, England and France were interested in making the most distant journeys possible, hoping to also discover new colonies. The Russian empress Catherine II hoped that all Russian astronomers would succeed within the Russian territory. Moreover, she also invited foreign scholars for the observation of this spectacle, so that a variegated society of Swiss, Swedish, French, and Russian scholars could meet.

Altogether the transit was observed by 151 observers at 77 stations distributed worldwide, but the result was still unsatisfactory. Although many other discoveries were made, the astronomical unit could by no means be derived with sufficient accuracy. Therefore, the two Venus transits of 1874 and 1882 were also of considerable scientific interest. In

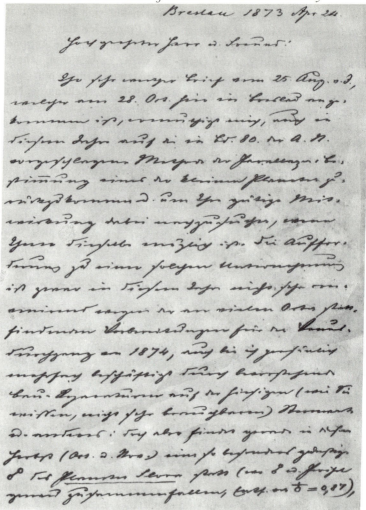

Fig. 8. First page of a letter by J. G. Galle to the American astronomer B. A. Gould in Cordoba (Argentina). This letter contains, among other things, the suggestion to observe the opposition of the minor planet Flora for the purpose of determining the solar parallax

1874 approximately fifty expeditions were outfitted, among them several German ones. One of the German expeditions which carried out its measurements in the southern hemisphere derived a value of 8".880 for the solar parallax from a total of 751 measurements. In spite of the accuracy achieved, the linear error always amounted to about 200 000 km.

Very accurate values for the solar parallax can also be obtained from parallax measurements of minor planets. J. G. Galle suggested this for the first time in 1872 (Fig. 8). At that time, however, all the known asteroids approached no closer to the Earth than 0.8 AU, never closer than Mars or

Venus at their closest approaches to the Earth. G. Witt successfully took up Galle's suggestion after the discovery of the asteroid Eros (1898), which approached the Earth within 0.15 AU. The case of the asteroid Amor, discovered in 1932, was even more favorable. With the application of astrophotography these celestial bodies represent ideal aids for the determination of the solar parallax. Spencer Jones and Rabe obtained the most accurate values of 8″.790 and 8″.78916 in 1930. Even the later radar echo measurements were only slightly more accurate. The astronomical unit was finally fixed by international definition, corresponding to a solar parallax of 8″.80.*

DOUBLE STARS

The discovery and investigation of double stars is one of the greatest achievements of classical astronomy.

A few closely-spaced pairs of stars had been known to astronomers for quite some time. However, a systematic observational search for these objects had not been carried out. Galileo had called for this, but the suitable instruments had not been built. However, in the last third of the eighteenth century Christian Mayer at the observatory in Mannheim ordered suitable telescopes and found with them a whole series of closely situated stars; at just about the same time N. Maskelyne in Greenwich obtained similar results. In a treatise published in 1778 Mayer came to the conclusion that these objects were planetary systems, but few believed this; only Lichtenberg gave his ideas some credence. Other contemporaries adhered to the notion that the closely situated stars had nothing to do with each other. They thought these stars appeared close together only by chance, and that in actuality they were at greatly different distances.

William Herschel also agreed with this assumption when he set out to use double stars for the measurement of stellar parallaxes. The faint neighbor of a brighter star would be used as a fixed reference point, for it was assumed that this star was much more distant than the brighter star, its parallax necessarily being negligibly small. Herschel intended to systematically survey the sky for suitable objects, and because he wished to measure the smallest angular changes, he developed a series of new technical aids specially for this purpose. Most importantly he improved the standard filar micrometer so that large separations could be measured along with the positions angles of the stellar pairs.

After several years he had discovered almost 1000 double stars, and from this he came to doubt the notion of the chance pairing of these objects. From observations of the position angles carried out over decades Herschel determined a motion about each other for several pairs of these objects. As a result it appeared certain that double stars were physically related objects.

* This corresponds to an Earth–Sun distance of 149.5 million km (92.9 million miles). (Tr. note)

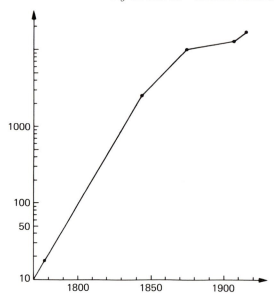

Fig. 9. The number of known double stars (1770–1915)

A further convincing proof for the physical relationship of numerous close pairs of stars was their common proper motions. Bessel and Piazzi confirmed that the parallel motion of the double star components of 61 Cygni was particularly apparent. When Bessel compared his observations of 1815 with observations that Bradley had made, he found a change of position angle of about 60 degrees and a common proper motion of 7 arc minutes. From this he concluded that the two stars jointly moved along in space and simultaneously traced out an orbital motion about a common center of gravity. The orbital period amounted to about 350 years. From Kepler's Third Law and the measured orbital period Bessel showed that the linear separation of the two components must amount to 50 AU. From a comparison of this value for the orbital size with the apparent angular separation Bessel derived a parallax for 61 Cygni on the order of $\frac{1}{3}$ arc second (the first determination of a dynamical parallax)[28] – a value which he actually later measured.*

---

* If a is the semi-major axis of the orbit of the secondary star about the primary, measured in seconds of arc, P is the orbital period in years, and M_1 and M_2 are the masses of the primary and secondary in solar masses, then the parallax π'', measured in seconds of arc (the reciprocal of the distance in parsecs) is

$$\pi'' = \frac{a''}{(M_1 + M_2)^{1/3} P^{2/3}}$$

Because the parallax depends on the cube root of the mass sum, one need not know this sum accurately, but one must still estimate it. (Here the linear orbit size, i.e., semi-major axis, measured in AU, is a''/π''). (Tr. note)

With the double stars a whole new class of celestial bodies was discovered which led to valuable results for astronomy and to important new insights concerning the laws of the cosmos.

The most successful discoverers of double stars right after William Herschel were his son John and the Russian astronomer Wilhelm Struve at Dorpat Observatory. John Herschel discovered many double stars in the southern hemisphere. However, Struve deserves credit for having carried out the observation of double stars with extraordinary precision. He used one of the best instruments available, a refractor built by J. Fraunhofer with a specially made positional filar micrometer. The excellent equatorial mounting, the modern micrometer, and the precise clock drive, which accurately drove the whole instrument according to the apparent movement of the stars, were prime reasons for Struve's success in the area of double star measurement. Except for Struve, no one at that time could make positional measurements accurate to a few hundredths of a second of arc. Thus Struve became the real founder of double star astronomy. In his extensive treatise *Stellarum Duplicium et Multiplicium Mensurae Micrometricae*, published in 1837, he reported on about 3000 systems, among which were 64 triples, 3 quadruples, and one quintuple system.*

The discovery of double stars and the observations of their motions immediately led astronomers to the question of the determination of the orbits of these objects. From this arose the second question of which law the motions of these objects follow. The Newtonian Law of Gravity had proven its marked applicability for objects within our solar system, but the double stars belonged to the stellar world and were situated incomparably further than the most distant planets of the solar system. Should the law also apply or does a completely different natural law rule there? If the motion of the double stars follows the Law of Gravity, the objects must move according to the Keplerian laws like the bodies of the solar system. Assuming the applicability of the Law of Gravity, F. Savary developed a procedure in 1827 for orbit determinations based on only a few observations. Because the results agreed with the observations it was evident for the first time that gravitation is a universal phenomenon which operates in the deepest cosmic realms.

The observations of double stars were carefully carried out in the following decades. Indeed, many of these objects move so slowly that the data required for an orbit determination were only obtainable with observations carried out over a very long period of time. During the nineteenth century many astronomers concerned themselves with the

---

* Baron Ercole Dembowski of Milan devoted the last thirty years of his life to the revision of the Dorpat catalog; he was a noted, as well as very active, double star observer. Two volumes of his observations were edited jointly by Otto Wilhelm Struve and the famous Italian astronomer Schiaparelli. (Tr. note)

determination of orbits of double stars, especially J. F. Encke (1832), J. Herschel (1833), E. F. W. Klinkerfues (1855), and C. Flammarion (1874).*

Bessel achieved a triumph of particular merit with the application of the Law of Gravity in the realm of stars. His many years of stellar observations had allowed him to show a noteworthy peculiarity in the motions of the stars Sirius and Procyon: namely, both stars showed oscillatory proper motions. From this Bessel concluded that the two bright stars must be components of double star systems whose fainter components manifest themselves solely through their gravitational effects – an 'astronomy of the invisible'. Later C. A. F. Peters attacked the problem more directly and derived the orbital elements of Sirius' invisible companion. In January 1862 A. Clark (USA) found the companion of Sirius (a star of magnitude 8.5) during a test of the optics of a new large refractor. The position agreed well with the prediction which Auwers had already published. In 1896 the companion of Procyon was found with the Lick refractor, for which Auwers had also determined the orbital elements. These faint companions are 'white dwarf' stars.

A tremendous extension of this 'astronomy of the invisible' was the tentative proof attained in 1963 of the existence of planets in a distant solar system. In this case the very same principle that Bessel had already used in the investigation of the Sirius and Procyon systems was used. From the extremely minute orbital perturbations of Barnard's Star (the greatest deviation was 0.015 arc seconds) P. van de Kamp demonstrated the existence of two invisible companions whose masses are comparable to the largest planetary masses in our solar system. Recently the existence of these planets has come to be regarded as uncertain; further measurements are required for a final decision.

With double star astronomy the study of the stars had opened up an extremely important chapter, for it was possible for the first time to determine stellar masses and to expand incredibly the fixed ideas concerning the distant stars. However, the study of double stars did not reach full efficacy until one could ascertain the dimensions of the orbits without parallax measurements, through the application of spectroscopy and the use of the Doppler effect (which gives absolute velocities). In this case it is clear that some areas of astronomy cannot be unambiguously categorized as classical positional astronomy or modern astrophysics. Astronomy and its greatest success are directly due to the assimilation of different research methods, as is shown particularly in our century.

* The most famous twentieth-century discoverer of visual double stars is S. W. Burnham. In 1900 he published a catalog of 1290 double stars discovered by him from 1871 to 1899. He also compiled data of others, and in 1906 published *A General Catalogue of Double Stars within 121° of the North Pole*, containing 13 665 pairs. Important work in the area of double stars was carried out at Yerkes Observatory and Lick Observatory. (Tr. note)

STELLAR DISTANCE MEASUREMENTS

The efforts of astronomers concerning the measurement of stellar distances exemplify the process of striving for ever better accuracy. Every prominent astronomer since Galileo has addressed himself to this challenge.

Many unintentional discoveries were made that held great significance for astronomy before the tiny shifts of stellar position resulting from the Earth's motion about the Sun (i.e., trigonometric stellar parallaxes) were actually measured.

As is well known, Copernicus did not doubt that the stars exhibited parallactic shifts of position, and he also considered the stars to be so distantly situated that their parallaxes could not be measured within the scope of the observational accuracy obtainable at that time. During Copernicus' time observational accuracy amounted to a few arc minutes. According to him, the inability to measure parallaxes meant that the distances of the stars must be greater than 1000 AU.

The greatest observer of the eighteenth century, the Englishman James Bradley, could measure positions to an accuracy of 0.5 arc second; however, he too could not observe a parallax. According to him the stars must reside in the incredible depths of space at a distance by no means less than 400 000 AU. Bradley's observational precision did allow him to discover two important effects – aberration and nutation.*

These two reduction elements, measured in the succeeding era by numerous researchers with increasing accuracy, were important prerequisites for the determination of more accurate positions, but the question of how much further observational accuracy had to be taken in order to also uncover a trace of the still smaller parallaxes remained. Bessel, the great master of positional astronomy, accurately described the unyielding, obstinate struggle to measure these tiny values when he wrote that the hope fostered by a sound belief in the existence of parallaxes would wither away if it could be proven 'that the trodden pathway of human artistic skill and human reasoning which had born no fruit was the ultimate limit attainable.'[27]

Bessel concerned himself quite extensively with the errors of measurement resulting from the observational instruments, on the assumption that if these errors could be reduced or measured, the result would be a greater level of precision without more advanced instruments. Accordingly, Bessel used proper motion values to index objects whose parallaxes should be

---

* Aberration is the apparent displacement of starlight due to the motion of the Earth (rotation about its axis and revolution about the Sun) and the finite speed of light. Nutation is a term relating to periodic variations of the wobble of the Earth's axis of rotation (i.e., changes in the precession). The largest nutational term results from the inclination of the Moon's orbit with respect to the ecliptic. (Tr. note)

investigated. He correctly assumed that stars with larger proper motions are on the average closer than the objects with smaller proper motions. By 1812, according to this criterion, he had selected 61 Cygni, and he wrote a detailed article about it. The large proper motion of this star allowed him to entertain the hope of measuring its parallax. Incidentally, from the apparent angular separation of the two components the orbit size for the two objects could be determined in absolute units, assuming the mass sum (see p. 42). 'One can therefore hope', wrote Bessel, 'that the astronomers with access to excellent equipment will study this noteworthy pair of stars with zeal.'[28] A quarter century later, however, no one had succeeded in measuring a parallax. Bessel himself had had 61 Cygni on his observing program since 1815. After the Fraunhofer heliometer was installed at the Königsberg Observatory in 1829 (Fig. 10), he saw new prospects for success. In 1837/38 he finally succeeded in freely and incontestably demonstrating the parallax of 61 Cygni. From hundreds of observations he derived the tiny angle of 0.3136 arc seconds as an average value for the parallactic shift; that corresponded to a distance of 10.3 light-years.

Immediately after this Wilhelm Struve in Dorpat published a successful parallax measurement. He had found a parallax for Vega in the constellation Lyra of 0.26 arc seconds, actually before Bessel had obtained his value. A year later the Scottish astronomer Henderson at the Cape of Good Hope found the parallax of 0.92 arc seconds for α Centauri, making it the closest star other than the Sun. Bessel's measurement turned out to be the most accurate of the three values; it was later revised only slightly.

The first parallactic measurements did not suddenly foster belief in the Copernican system, which had been little doubted for a long time. These measurements were, however, of great fundamental significance; with them astronomy ventured into greater depths of space than ever before. The parallax measurements fostered the expectation that the actual distribution of stars in space could be determined, laying the basis for the kind of scientifically-founded investigation of the structure of the universe William Herschel had suggested. Bessel stressed another important theoretical–conceptual benefit of the first parallax measurements: they demonstrated that a scientific goal proposed by many could be attained.

Given the methods used by Bessel and his colleagues, the measurements were extremely involved; this is the reason why it was a long time after the first parallax measurements before other distance measurements were made. Not until the introduction of photography did a fundamental change come about. In addition to increasing the simplicity of measurement, the application of photography improved accuracy, for now one could determine the position of the star of interest with respect to many faint stars. In 1900 only about 150 stellar parallaxes were known. By 1924 1500 parallaxes were derived, primarily through the untiring work of D. Gill and, above all, F. Schlesinger. The parallax catalog which Louise F. Jenkins and

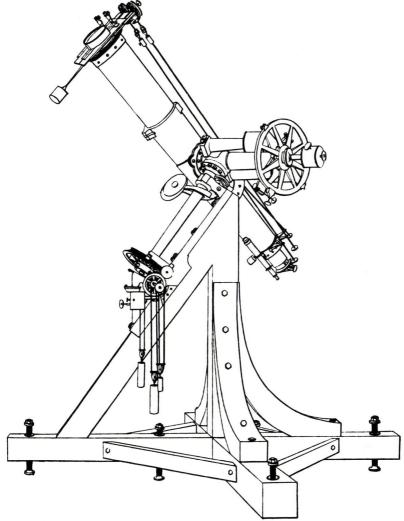

Fig. 10. Heliometer of J. Fraunhofer, with which Bessel measured the parallax of the star 61 Cygni

Schlesinger published in 1935 contained the parallaxes of about 10 000 stars. Today parallax programs are being carried out by different observatories with the help of long-focus refractors, or, recently, special reflectors (US Naval Observatory, Turin Observatory), which allow the measurement of fainter objects.

The application of astrophysical methods considerably increased the number of procedures available for the determination of stellar distances; this was especially important because the trigonometric parallaxes were limited to a certain distance range (about 100 light-years).

In 1908 the American astronomer L. Boss first showed that the distance of a star cluster could be determined with great accuracy using the 'moving cluster method'. In this method the tangential velocity components of the cluster members (i.e., their proper motions) were measured and combined with the radial components obtained from the Doppler shifts. The proper motions can be converted to linear units, from which the distance follows. In a similar manner the linear dimensions of a system can be found from double star observations, for which spectroscopic orbit determinations were used, and these dimensions can be used to determine distance.

Finally, the method of spectroscopic parallaxes, developed in 1914 by W. S. Adams and E. Kohlschütter, is of particular importance. It was based on spectral criteria for the absolute magnitudes of the stars, which then could be used for a derivation of stellar distances. These and other procedures again point out the relationship between the methods of astrophysics and positional astronomy. Questions which at first belonged only to classical astronomy were later solved by the substantial application of astrophysics. Similarly, astrophysical methods are based in part on the achievements of positional astronomy.

ASTRONOMICAL GEOGRAPHY AND GEODESY

The extraordinarily rapid growth of celestial mechanics and positional astronomy from about 1750 to 1850 is not comprehensible solely from the scientific motives already delineated. This rapid progress is closely bound up with the fact that the astronomical results which were required for the calculation of orbits of celestial bodies and for their precise observation could also be used for extremely important practical purposes – the determination of geographical positions on the Earth and the investigation of the shape of our planet.

The determination of geographical latitude depends essentially on an angular measurement, namely the determination of the elevation angle of the celestial pole at a given place. The determination of longitude, however, is more complicated. In order to obtain the longitude difference of one place from another situated westward or eastward from it a time difference must be measured according to the apparent daily motion of the stars. In the case of the Sun the observers set their clocks to 12 o'clock noon at the moment of the maximum elevation angle of the Sun. The problem lies in having to compare the two clocks with each other in order to use the positional time difference to derive the longitude difference of the places. The clock comparison could only directly be obtained by the transport of the clocks to a common place. Accurate chronometers which are allowed to change their rates only in a precisely known way during the transport are required for this. Both observers can also observe certain celestial phenomena which are predicted in advance to take place at a precisely determined time. Because each of the observers measures this moment according to his clock, the difference of the readings corresponds likewise to the longitude difference.

Though these procedures are easy in principle, they are quite complicated in practice. The solution of this problem required the sagacity and perseverance of numerous astronomers, mathematicians, and instrument makers over many decades. The results obtained through this work greatly superseded the original purpose and gave all of astronomy substantial and progressive impetus. An example of how deeply longitude measurement influenced other astronomical problems was given in the case of a unique method which had been developed in 1514 by J. Werner. This method of lunar distances requires that the angular distance of the Moon from specific fixed stars or planets be measured at particular times at a position of known longitude. From the determination of the value of the point in time of this lunar position one can immediately determine the longitude of another place from the time difference. If one desires to successfully use this procedure, however, it is necessary that one be sufficiently precise in measuring the motion of the Moon and that one measure its position against the background of stars with sufficient precision. In the seventeenth century the longitude determinations made with this method were seldom more accurate than 5 degrees. For positions on the Earth near the equator this amounted to errors on the order of a few hundred kilometers.

As trade and shipping developed the problem of longitude determination became of paramount importance for the unfolding of capitalist nations. Inadequate determinations of position cost thousands of sailors' lives, and the number of naval casualties became of dire concern. During the seventeenth century a ship that returned home unscathed from a trip to China was reason for great celebration throughout the whole country. It is understandable that the states with modern production and developed trade did everything possible to improve the astronomical foundations of navigation. As a result, hardly any other branch of astronomy was carried on with such an expenditure of brainpower, energy, and money as was the surveying of the Earth. These investigations were without exception amply state-funded and consequently made rapid progress. The inner logic of astronomy addressed itself at the same time to this problem as the order of the day, and this accelerated the development.

In 1598 Philip II of Spain set aside a significant sum for the solution of the problem of determining position at sea. A century later The Netherlands followed suit. The best example of the fruitful interaction between astronomy and practical, social requirements is the history of the founding of the famous Greenwich Observatory. Charles II of England had an interest in possibly determining geographical longitude at sea from the knowledge of the position of the Moon against the background of fixed stars, and he set up a commission to examine the existing alternatives. When questioned, the 28 year-old J. Flamsteed expressed the opinion that if this method were to succeed, the positions of fixed stars as well as the Moon's motion must be determined with substantially increased precision. Subsequently, Charles II gave the order for the founding of Greenwich Observatory and appointed Flamsteed director. The research program was strictly laid out – the most accurate tables of the motion of celestial bodies, especially the Moon, and the most precise catalogs of stellar positions were to be made.

Precise determinations of position based on astronomical measurements were also an extremely topical problem for cartography. In 1752 only 139 places on the Earth's surface had had their positions accurately measured.

The English Parliament announced a well-publicized reward for the working out of a method of determining longitude at sea to within $0°.5$ accuracy. This governmental decree to science led to extremely important results; it stimulated both theoretical investigations and practical inventiveness. The subsequently famous chronometer of the English clock-maker Harrison (1761) ran false by only five seconds during an ocean voyage of 161 days.

T. Mayer led the theoretical efforts concerning the solution of this problem. Mayer, like L. Euler, described the theory of the Moon's motion; the lunar tables worked out by him, his primary result, were handed over to the Admiralty in 1755. After 1757 the English Admiral Campbell became the first seafarer to determine geographic longitude at sea with the use of a Hadley sextant and the lunar positions calculated by J. Bradley and T. Mayer. With the *Nautical Almanac* (1766) N. Maskelyne created the standard ephemeris for the convenient, practical application of the theoretical results.

The convincing success of longitude determination by means of lunar distances encouraged the further pursuit of the complicated celestial mechanical problem of lunar motion. Once again publicly announced prizes consciously fostered this development. For example, in 1800 the French Bureau des Longitudes called for the determination of the values of specific coefficients of the equation of motion of the Moon from the comparison of good lunar observations and the derivation of new tables. These were to be derived 'with a convenience and confidence sufficient for the calculation'.[29] The high points of subsequent research in this area were the later calculational simplifications of Bessel (1832), the substantially improved lunar tables of P. A. Hansen (1857), the critical and comprehensive documentation of all historical eclipse observations by S. Newcomb (1878), and the modern lunar theory of E. W. Brown (1896).

Besides the method of lunar distances there is another whole series of procedures which lays great claims to observational accuracy and theory and which incessantly fostered further development. One writer described the importance of such investigations in this way: 'It is sufficient to say that . . . those expenses might have been nothing more than a waste of money, useless for capitalist society. How great the influx of good maps was and

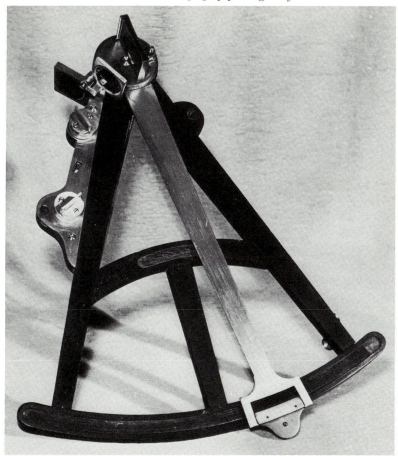

Fig. 11. Reflecting sextant by Fiedler (without telescope)

how important these were in regard to politics, statistics, and military matters – there can now be only one viewpoint on this matter.'[30]

At the turn of the nineteenth century approximately 50 per cent of all active astronomers (as well as others) were involved with the problems of geographical position measurements. The first astronomical journals were at the same time focal points for the publication of astronomical–geographical research. During the first years and decades of the nineteenth century the quantity of geographical publications in them was greater than the number of specifically astronomical works. The *Monatliche Correspondenz zur Beförderung der Erd- und Himmels-Kunde* published about 3000 geographical position measurements in its first fourteen years.

Astronomical geography came in direct contact with another important problem in the surveying of the Earth, namely the question of the shape of our planet.

From the observation of flattening of other planets, especially Jupiter, it was already assumed that the Earth was not a sphere. If the Earth should be similarly flattened at the poles, two places near the pole on the same meridian of longitude on the Earth with a latitude difference of one degree must be situated further apart than a corresponding pair of places at the equator. Astronomical angular measurements (i.e., longitude and latitude determinations) as well as distance measurements on the Earth are required to test this assumption. The first investigation of this with a claim to scientific precision was carried out by the French Academy of Sciences during the years 1735–1744. The required measurements were made in Lapland and Peru, and the results unequivocally demonstrated that the Earth was flattened at the poles.

The accuracy of these determinations still left a lot to be desired. The lack of an internationally accepted unit of length was also a serious problem, especially for the comparision of results. At the very least a standard length had to be strictly defined in terms of other units of length. A uniform system of measure had been an urgent need for some time, since the lack of uniform weights and measures also hindered commerce and trade. The overdue resolution of this question came about at the time of the French Revolution under the auspices of noteworthy scholars like Delambre, Coulomb, Lagrange, Laplace, and Monge. A wide-reaching reform of the system of measure was introduced, which was considered a child of the Revolution. It called for the adoption of length and weight standards and a calendar reform, as well as the general adoption of the decimal system. Because of political reasons, different regimes resisted the innovations originating from France, so it was necessary to choose a natural unit of length that could be derived from the dimensions of the Earth. The meter was defined to be one ten-millionth part of an Earth quadrant (i.e., a quarter of the Earth's circumference along a meridian of longitude). For this purpose a great meridian measure was started under the direction of J.-C. Borda. The angle from Dunkirk on the northern French coast to Formentera in the western Mediterranean Sea ($12°.4$) was measured along the Parisian meridian. This measurement had only historical significance, for although more substantial and more accurate measurements were achieved later, the unit of length was nonetheless retained.

The surveys of meridian and latitude arcs were carried out with great pomp on an international scale. In England and Scotland, in northern Italy, Denmark, and Holstein expensive and ever more accurate measurements were carried out. In Russia, Wilhelm Struve began the so-called Livonian degree measurement, carried out from Lithuania to the Turkish border, for which Swedish scholars took care of the northern measurements.

The Prussian degree measurement of F. W. Bessel was of particular note in the nineteenth century, since it exhibited the greatest precision. After thirty years of measurements which Bessel procured with J. J. Bayer, the reduction of the data took several additional years, and the results were not published until 1838. A distinct scientific discipline, geodesy, gradually arose from astronomy and the astronomical geography connected with it.

Other important geodetic projects were carried out during the nineteenth century; most importantly, an international Earth survey program came about in 1862 from the ideas of Wilhelm Struve and J. J. Bayer. The Geodetic Institute in Potsdam, originated through Bayer's initiative, played a coordinating role in the program.

In the first half of the nineteenth century research into the question of the Earth's shape had already resulted in a quite accurate value for its oblateness: 1/298, i.e., the equatorial diameter of the Earth was shown to be 1/298 greater than the polar diameter.[31] However, it had also been unequivocally shown that the Earth is not an ideal ellipsoid but exhibits an irregular shape, caused by the uneven mass distribution in the Earth's interior, which is totally independent of the positional differences of altitude. This considerably complicated the attempts at measuring the Earth and contributed to the fact that these problems no longer lay in the realms of astronomical research. Determining the irregular shape of the Earth brought about the necessity of an international cooperative venture, because the ever more precise measurements of smaller arcs said nothing conclusive about the whole Earth.

Although geodesy eventually came into its own, astronomy is to be thanked to a great degree for practical advances for a great number of problems. This was not the last time astronomy provided the impetus for greatly increased measurement accuracies. Astronomical precision allowed instrumental observational technique, instrumental theory, and celestial mechanics to be raised to greater heights.

A prime example of this precision is the discovery of the oscillations of the pole. At the beginning of the nineteenth century S.-D. Poisson and F. W. Bessel discussed the supposition on theoretical grounds that the elevation of the pole should not be constant. In the 1840s Bessel had even hoped to measure such tiny oscillations, fractions of a second of arc. He was right – in 1888 F. Küstner discovered the very minute oscillations of the pole with sufficient accuracy.* This discovery resulted in the establishment of the International Latitude Service.

---

* It was Leonhard Euler, the famous Swiss mathematician, who predicted in 1765 that there should be a 10 month period of latitude variation. Küstner was the first to demonstrate the effect of latitude variation, but he found it accidentally. The American S. C. Chandler is the principal name on this topic. He found that there are two oscillations (neither with a period of 10 months): a 14.2 month period, now called the Chandler component, is a free oscillation arising from the spheroidal shape of the Earth; a 12.0 month component is a forced oscillation due to meteorological effects of annual period such as seasonal changes in air masses in the northern and southern hemispheres. See W. Markowitz's review in *Sky and Telescope*, August 1976, p. 99. (Tr. note)

PLANETARY RESEARCH

The experts in celestial mechanics were not the only scientists interested in the planets; an active interest grew concerning the nature of these bodies and especially concerning their surface features.

Planetary research was given a particularly strong impetus by the rapid development of the telescope after the end of the eighteenth century. The research of this epoch, which later received new stimulations through the application of astrophysical methods, emphasized the relative and historical character of scientific research. Only a fraction of the questions posed could be definitely solved, many new riddles emerged, and numerous questions required space travel for their final resolution. Among these questions is that of the origin and development of the planetary system. The British Astronomical Association (founded 1890) made important contributions to planetary research; it organized special sections for the promotion of this branch of astronomy. Scientists such as T. E. R. Phillips, E. W. Maunder, W. T. Lynn, and A. de la Cherois Crommelin worked with the BAA.

A review of planetary research from the nineteenth century to the first decades of the twentieth century leads to the disappointing conclusion that knowledge about the nature of the planets remained relatively limited in spite of intensive endeavours. To a large degree astrophysical methods did not achieve the same level of success in planetary astronomy as they did in stellar physics (see pp. 69 ff.). It is a curious fact that our closest cosmic neighbors, the planets of the solar system, are in no way the most well understood objects.

Although planetary astronomy was not a prime area of research in the nineteenth century, the public became particularly interested in this branch of astronomy. The generally intriguing question of the existence of life forms in the universe, which is associated with planetary research, is not the only reason for this interest.

The inferior planets

The first extensive investigation concerning the physical nature of the planet Mercury was carried out by J. H. Schroeter, who in 1800 and 1816 published two copious volumes, his *Mercurian Fragments*. The discussion of the rotational period of the planet, which Schroeter determined to be about 24 hours, was one of the most important results. In addition to this, from the jagged appearance of the border between light and dark on the planet (the terminator), Schroeter concluded that this planet must be covered with mountainous areas.

After the investigations of Schroeter it was a long time before anyone else tackled the difficult observational problem of the little planet Mercury.

Fig. 12. Schroeter's great reflecting telescope in Lilienthal

Schroeter's book remained the standard work until J. K. F. Zöllner obtained new results concerning Mercury in 1874, in connection with his photometric investigations. He studied the light reflectivity (albedo) of the surface and drew the conclusion that there is a far-reaching similarity between the surface of Mercury and the Earth's moon – a result completely confirmed in early 1974 by the American probe Mariner 10. In 1889 the Italian planetary observer Schiaparelli amended the value of the rotational period given by Schroeter and others and expressed the notion that the planet exhibited a bound rotation, i.e., it turns about its axis in the same time it takes to revolve around the Sun. Schiaparelli drew surface maps of Mercury, which he painstakingly put together from many hundreds of individual observations. The nature of the suspected surface details remained essentially undecided. In 1832 Bessel had already determined the diameter of Mercury to be 4855 km, quite close to the value of 4840 km accepted today. Because Mercury has no moons, the determination of its mass also manifested considerable difficulties, for it could only be derived from the orbit perturbations by Mercury on comets or the neighboring planet Venus. In 1835 Encke measured its mass from the perturbations on the short-period comet named after him; his value was 1/4 686 571 of the Sun's mass (modern value: 1/5 970 000).

The planet Venus also created difficulties for observers from the start, for it exhibits no surface details whatsoever. By the middle of the eighteenth century the great Russian scientist M. V. Lomonosov had discovered that this planet must be surrounded by an atmosphere. On the occasion of the

Venus transit of 1761 it had occurred to him that the dark planetary disk in front of the Sun was surrounded by a bright rim of light, which he took to be the atmosphere of the planet. The American astronomer D. Rittenhouse confirmed this; he observed the same phenomenon in 1769 during the second Venus transit of the eighteenth century. Both reports, however, were not published until much later, long after Herschel's similar observations were generally known, so that Herschel is often credited today with the discovery of this effect.

The lack of surface details made a determination of the rotational period of the planet impossible, and there were contradictory statements about it. Toward the end of the nineteenth century Schiaparelli spoke of a bound rotation; other astronomers derived a shorter rotational period from the motion of surface details which later proved to be optical illusions. The inclination of its rotational axis was also unknown.

In order to study more accurately the effect of optical illusions, W. Villiger published in 1897 a series of simulation investigations in which Venus was represented by small balls observed and drawn from various distances.

The hypotheses concerning the nature of Venus' surface all relied upon very sparse observational material and were therefore very vague. Whether the planet was principally covered by bodies of water or whether there were giant marshy continents could not be scientifically decided. New hypotheses did not come about until the investigation of the spectrum of the planet. The Venusian atmosphere was discovered to be mainly carbon dioxide, so one must infer the operation of the so-called greenhouse effect – the Venusian surface must be hot enough to turn all water to steam.

In spite of the fact that the spectroscopic measurements were improved and also allowed determinations of temperature, knowledge about the planet Venus remained full of riddles. The astronomers reached the limits of earthbound research, even though this very bright planet was so near. Real breakthroughs in Venusian research were not made until the landing of the Soviet probes in the second half of the twentieth century.

The superior planets

Mars has always been one of the most interesting planets because of its fiery red color and its widely varying luminosity.

In contrast to Mercury and Venus, Mars has surface details visible to observers with small telescopes which can be used for a determination of the planet's rotational period. The investigation of Mars' surface nature was quite successful; half a century after the invention of the telescope Huygens had already derived an accurate value of the rotational period (24 hours). He also discovered the large dark formation which is known today as Syrtis Major. On the occasions of the Mars oppositions of 1777 and 1779 Herschel found a rotational period of 24^h 39^m $21^s.7$, only two minutes different from the modern value. He referred to Syrtis Major as the 'Hourglass Sea', not only because its shape resembles an hourglass, but also because it had been used for time reckoning.

Herschel also was the first to observe the seasonal color variations on Mars; he deduced the existence of an atmosphere from the appearance of the so-called polar caps – white patches at the north and south poles of the planet which he took to be masses of snow and ice. From the inclination of Mars' rotational axis, which Herschel determined to be 23° 13', he concluded that there were wide-ranging similarities between Mars and Earth. The interpretation of these similarities was not refuted until a few years ago by the flybys of American and Soviet satellites.

W. Beer and J. H. Mädler, who drew the first Mars maps, were notable observers of Mars. A. Secchi, J. N. Lockyer, R. A. Proctor and others followed with subsequent maps.

The most breathtaking results of Mars research in the nineteenth century, which influenced all subsequent investigations, were obtained by Schiaparelli during the favorable Mars approach of 1877. Schiaparelli found a series of finely arranged, straight formations on the surface of the planet, which he termed *canali*. He published his investigation in an Italian academic journal which was usually known only to those in the technical world. This time, however, it was different. The results soon became the topic of the day, and the contents of Schiaparelli's article were published in simplified form in many other places. Numerous astronomers were also fascinated by a thought Schiaparelli himself had not professed; they assumed that the *canali* (literally, 'channels') were to be regarded as ingenious structural works of some kind, built by intelligent Martians who wished to channel the melting ice masses of the poles to the presumably dry central regions of the planet. The famous French astronomer and author Flammarion wrote in one of his books that one should feel sorry for anyone who had not seen Mars in a telescope, for inspiring knowledge of a strange world in the vast celestial realms, with its continents, seas, islands, and shores, would remain unknown to them. The Earth is only one little corner of the universe, he continued, and we might be able to visit other ports in space which are inhabited by extraterrestrial beings.

A number of facts did not fit well with the picture of the Martian canals. Their implied dimensions had to be considered unrealistically large; several were many kilometers wide and up to 1000 km long. Even with the most liberal estimate of the amounts of water to be transported the canals were much too large. Still more contradictory was the report that some canals changed their shape; many specific canals were not seen at all by some observers, while these same observers thought they saw other canals. Some canals were unequivocally observed, and this gave rise to an ever expanding literature concerning Mars which described intelligent Martians. Some scientists seriously believed in life on Mars. The American astronomer Percival Lowell was one of the most prominent of these advocates at the close of the nineteenth century.

In the same year during which Schiaparelli discovered his sensation-causing *canali*, the American A. Hall found the two moons of Mars. He had access to the most powerful telescope in the world at that time, with an objective diameter of 65 cm. He gave them the names of the two steeds which in Greek mythology drew the chariot of Ares (Mars): Phobos (fear) and Deimos (dread) – suitable names for the planet of war.

In 1909, when spectral analysis was advanced enough for planetary research, G. A. Tikhov, a well-known Russian observer of Mars, had a simple but extremely interesting idea. He began to compare the spectra and optical peculiarities of certain regions of the Martian surface with those of selected terrestrial vegetation zones. He became as a result of this work the founder of astrobotany. Tikhov and his collaborators felt they had proven that vegetation exists on Mars which is related to the flora of subarctic terrestrial regions, but these viewpoints were disputed. On 20 July and 3 September 1976, respectively, the Americans landed the probes Viking 1 and Viking 2 on Mars and began the work of searching for evidence of life. The results of the experiments carried out are still being disputed, but on the whole they indicate no life on Mars. The final clarification of these questions must be left for subsequent space probes. However, it already seems certain that at best we would be dealing with the most simple forms of life.

The degree to which the use of space probes enlarged and enriched the knowledge of the nature of the planets is especially clear in the results of research concerning Mars. One of the biggest surprises in the history of modern planetary research was in 1964 when numerous craters were seen on the photographs of Mars taken by the probe Mariner 4, so that the planet regarded until then as similar to Earth now had to be considered a cosmic 'brother' of the Moon instead. New results have further modified this picture, and, taking into consideration the smaller mass and surface area of Mars, a number of analogies to the Earth cannot be overlooked.

Although the great planet Jupiter, which follows Mars in order of distance from the Sun, is at a much greater distance from the Earth, its exceptional size alleviated some problems for planetary observers.

The first telescopic observers saw a great amount of surface detail on Jupiter. They took note of the remarkable flattening of the planet, and were also fascinated by the alternating configurations of the four large moons, discovered by Galileo in 1610. In the seventeenth century G. D. Cassini made the first determination of the rotational period from the periodic disappearances and reappearances of certain markings. The value of $9^h 56^m$ obtained by him was confirmed by other observers. The prevalent opinion was that the striated formations running parallel to the equator, as well as other markings, were atmospheric phenomena and not markings on a solid surface of the planet. In 1779 Herschel confirmed this for the first time when he found systematic variations in the specific rotational periods of these formations. He came to the conclusion that a particular spot in the Jovian atmosphere observed by him was rapidly running ahead. J. H. Schroeter also thought it was likely that the markings represented an analogy to our clouds, and he therefore doubted that these formations would suffice to determine the rotational period of the planet with sufficient accuracy. Actually, the determinations of the rotational period by Cassini, Herschel and others were so accurate that until now no greatly different values have been measured.

Bessel discovered one interesting fact concerning the planet Jupiter. From the periods of revolution of Jupiter's moons he derived the mass of the planet and found a value of 336 Earth masses (modern value 317.8); however, this value contrasted with a volume which amounted to 1500 times as large as the Earth's. The resulting average density of 1.3 g/cm^3 was remarkably accurate and gave rise to consideration of the inner structure of the gigantic planet.

Spectroscopy eventually gave us some results which served to bring attention to significant characteristics of the strange, light, giant planet. H. C. Vogel (1874) and H. Draper (1880) both found a dark band at the red end of Jupiter's reflected solar spectrum, and from this deduced a considerable effective temperature for the planet. The discovery made by E. W. L. Tempel in August 1878 was completely in line with this interpretation; on the surface of the planet he saw an oval, reddish cloud of enormous size. Tempel estimated the length to be one third of the Jovian diameter. The Russian observer F. A. Bredikhin, O. Lohse in Potsdam, and others dedicated themselves to further careful observations of the object. The generally accepted interpretation was that the red cloud was created by a gigantic lava flow reflected by the cloud layers floating above. The review of older Jovian observations then showed that the Great Red Spot had been seen in the seventeenth century by various observers, and that the spot had shown variable rotational speed. This result was not in accord with the accepted theory of the spot. When Donald H. Menzel determined the temperature of the planet to be −110°C the statements concerning the planet were even more contradictory.

The identification of spectral lines in the Jovian atmosphere at the beginning of the 1930s led to the realization that methane and ammonia were to be found there; these gases were found in more than traces, while no proof could be found for the assumption that hydrogen and helium were the principal constituents.

In 1938 R. Wildt put together a theory of the structure of Jupiter in which he attempted to account consistently for the total mass and the low density, as well as other details borne out by research. He assumed that the core of the planet was very dense, but not very extended, and was enveloped by an ice mantle which was itself surrounded by a shell of frozen hydrogen;

the observable atmosphere lies above this. This interpretation led again to new complications, because the turbulence in the atmosphere of the planet could be accounted for only with great difficulty. The new results of the American planetary probes which passed by the planets at relatively small distances led to the conclusion that Jupiter is a huge ball of hydrogen which has a core of heavier elements in its interior.

Astronomy encountered quite similar problems and results with the planet Saturn, which in several ways is closely related to Jupiter. The density of Saturn also turned out to be very small in comparison to the planets Mercury, Venus, Mars, and Earth (modern value for Saturn – 0.7 g/cm^3). Early observers like Grimaldi (1645) and Herschel (1789) found the value of 1/11 for the oblateness of this second largest planet; for Jupiter the value is 1/15.

Understandably, the ring system of Saturn excited great interest, because it was the only such formation known in the solar system. The existence of this ring system was first established in 1659 by C. Huygens, although Galileo had already discovered two 'knobs' one on each side of the planet, which he could not understand owing to the limited resolving power of his telescope. In 1675 G. D. Cassini found the division of the ring system named after him. He was the first to advocate the idea that the ring was made up of numerous small particles.

Toward the end of the eighteenth century William Herschel dedicated himself to an extensive series of observations of Saturn, its surface features, its moons, and above all the ring system. He took the surface features to be atmospheric phenomena like those on Jupiter. The first measurement of the rotational period of this giant planet, made by Herschel in 1793/94 (10^h 16^m 44^s), differed by only a few minutes from the modern accepted value.

Herschel had an interpretation of the ring different from Cassini's; he considered it to be a solid body, with the gap being a dark part of the ring. Later, he was of the opinion that there were two different rings, which revolved about the planet according to the Keplerian laws.

In 1785 the attention of astronomers intensified when Laplace published a mathematical investigation of the ring, in which the existence of more gaps was predicted. (Kant had already written about this in 1755.) J. H. Schroeter, S. H. Schwabe, W. C. Bond and his son G. P. Bond, J. F. Encke, and others observed Saturn with better and better instruments. In 1837 Encke found the portion of the ring system named after him. In 1850 G. P. Bond discovered a third ring, the so-called Crêpe Ring of Saturn, although it had been suspected earlier by J. G. Galle and had remained unconfirmed. This ring, the innermost of the three ring forms, later became known as the C-Ring.*

---

* Pictures sent back to Earth from Voyagers 1 and 2, which passed Saturn in November 1980 and August 1981, respectively, reveal the ring system to be made of hundreds, perhaps thousands, of little ringlets. (Tr. note)

Astronomers went to great lengths to explain the existence of Saturn's rings. In 1857 J. C. Maxwell wrote a work on the rings in which he came to the correct conclusion that the rings consist of a multitude of small particles. The American astronomer J. E. Keeler was able to confirm this convincingly well with the help of spectroscopic investigations, in which he measured the velocities of the different parts of the rings from the Doppler shifts and thereby found that the outer ring portions move more slowly than the inner ones, in accord with Kepler's Third Law.

The existence of the gaps can be explained by celestial mechanics. Particles in the gaps are perturbed by many moons of Saturn or something of the kind, such that they are excluded from certain orbital distances. Similar arguments were encountered regarding Encke's division, as D. Kirkwood proved, although this gap is not completely particle-free; rather, it is greatly emptied of particles.

The theory of the inner structure of Saturn developed quite analogously to Jupiter.

The first investigations of the planet Uranus were carried out by its discoverer, Herschel. Of the six moons which he believed he had found only two were confirmed; the other objects were faint stars which were situated in the same region of the sky. Herschel found a notable flattening of the planet, from which he deduced that Uranus rotates rapidly. However, subsequent observers disputed the value of oblateness. The reason for the contradictory statements concerned the inclination of the rotational axis to the plane of the orbit. It was unique in the solar system. The subsequent observations showed that the axis is directed nearly in the orbital plane of the planet, so that the planet rolls along in its orbit like a wheel going around a circular track.

The determination of the rotational period manifested the greatest difficulties, for surface details could only occasionally be observed. W. Buffham derived a rotational period of 12 hours from observations of 1870 and 1872 (modern value – 10^h 49^m). The physical state of the planet eluded detailed inquiry and it is not unfair to say that we know almost as little today about Uranus as we did 150 years ago. The spectroscopic investigations only demonstrated the presence of a lot of methane in the atmosphere, as had also been found for Jupiter and Saturn. From the mass and diameter resulted a value of density in between those of Jupiter and Saturn, only slightly greater than that of water. Similar models were sketched out for this planet like those which had been derived for the two giants situated closer to the Earth.

The discovery of a ring system around the planet Uranus was very surprising. On 10 March 1977 Uranus occulted the star SAO 158687 in the constellation Libra (magnitude $8^m.8$). Approximately 40 minutes before this star was hidden by the planetary disk several fadings of the star were noted which indicated the existence of five rings. A direct visual observation of the

Table 1. *Discovery of planets and their classical moons*

Name	Discoverer	Year
Uranus	William Herschel	1781
Moons of Uranus		
Titania	William Herschel	1787
Oberon	William Herschel	1787
Ariel	W. Lassell	1851
Umbriel	W. Lassell	1851
Miranda	G. P. Kuiper	1948
Neptune	J. C. Adams and U. J. J. Leverrier/ J. G. Galle	1846
Moons of Neptune		
Triton	W. Lassell	1846
Nereid	G. P. Kuiper	1949
Pluto	C. W. Tombaugh	1930
Moon of Pluto		
Charon	J. W. Christy	1978
Moons of Mars		
Phobos	A. Hall	1877
Deimos	A. Hall	1877
Moons of Jupiter		
Io	G. Galilei/S. Marius	1609/10
Europa	G. Galilei/S. Marius	1609/10
Ganymede	G. Galilei/S. Marius	1609/10
Callisto	G. Galilei/S. Marius	1609/10
V (Amalthea)	E. E. Barnard	1892
VI (Himalia)	C. D. Perrine	1904
VII (Elara)	C. D. Perrine	1905
VIII (Pasiphae)	P. Melotte	1908
IX (Sinope)	S. B. Nicholson	1914
X (Lysithea)	S. B. Nicholson	1938
XI (Carme)	S. B. Nicholson	1938
XII (Ananke)	S. B. Nicholson	1951
XIII (Leda)	C. T. Kowal and co-workers	1974
XIV (unconfirmed)	C. T. Kowal	1976
Moons of Saturn		
Mimas	William Herschel	1789
Enceladus	William Herschel	1789
Tethys	G. D. Cassini	1684
Dione	G. D. Cassini	1684
Rhea	G. D. Cassini	1672
Titan	C. Huygens	1655
Hyperion	W. C. Bond	1848
Iapetus	G. D. Cassini	1671
Phoebe	E. C. Pickering	1898
Janus*	A. Dollfus	1966

* Janus has not been confirmed. The Voyager pictures indicate that the Janus observations must have been of two co-orbital satellites, S 10 and S 11 (1966 S 1 and 1966 S 2).

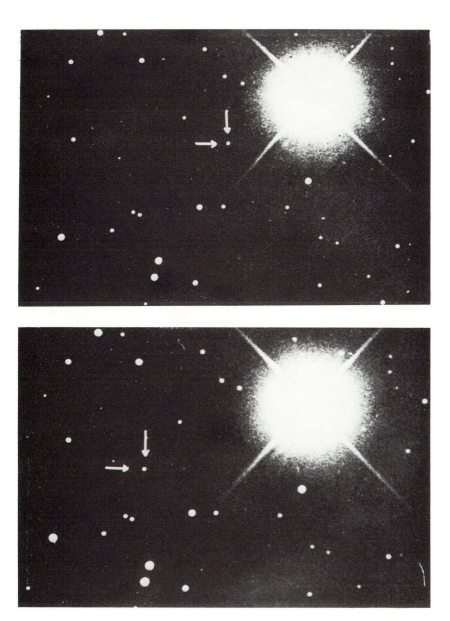

Fig. 13. Discovery photographs of the planet Pluto in the constellation Gemini (a) on 2 March 1930, (b) on 5 March 1930. The object marked by arrows is Pluto

ring system is not possible, owing to the faintness of the rings and their small distance from the planet.*

Similarly, the great distances of the planets Neptune and Pluto have so far prevented accurate understanding of the physics of these bodies. Only the data derivable with the methods of celestial mechanics can be obtained with some precision. One discovery deserves mention, however. A moon of the planet Pluto was discovered in 1978 in a routine investigation of photographic plates. On the otherwise circular image of Pluto the moon was made evident as a deformation. This deformation was also confirmed with previously taken photographs. The physical data of the satellite are not yet accurately determined.

MAPPING THE MOON

Detailed investigations of the Moon's surface began with the invention of the telescope. With his *Selenographia* (1661) J. Hevelius created the first detailed lunar map derived from careful telescopic observations. Hevelius also gave names to the different lunar formations, which, however, were soon superseded by the nomenclature of Riccioli. The latter made the Moon into an 'astronomers' cemetery', in which he named the objects on the Moon's surface after famous astronomers, a tradition which has been upheld ever since; in particular, the objects on the back side of the Moon remained completely nameless until their first observation by the Soviet probe Lunik 3 (1959).

T. Mayer issued in a new era of lunar cartography in 1750. He measured the lunar surface not by eye, but on the basis of micrometric measurements of different surface features. However, the large lunar map that he planned was not finished. A small map with a 20-cm diameter, which G. C. Lichtenberg published for the first time in 1775, was for many decades the height of achievement (Fig. 14).

Toward the end of the eighteenth century J. H. Schroeter, who had better instruments to use, improved upon Mayer's achievements. Schroeter did not plan any kind of comprehensive lunar map, but proceeded with investigations of detail with precision never before attained and made accurate descriptions of the lunar formations. He also concerned himself with the measurement of the diameters of numerous craters and the altitudes of many places. He published his results in two volumes, his *Selenotopographical Fragments* (1791 and 1802; Fig. 15). Concerning the *Fragments*, G. C. Lichtenberg wrote to Schroeter with unreserved enthusiasm: 'You have certainly carved a monument which will be as

---

* The rings of Uranus have been detected by means of ground-based near-infrared observations.

    We now know that Uranus and Saturn are not the only planets in the solar system to have rings. In 1979 the American probes Voyager 1 and Voyager 2 sent us pictures of a ring system around Jupiter. (Tr. note)

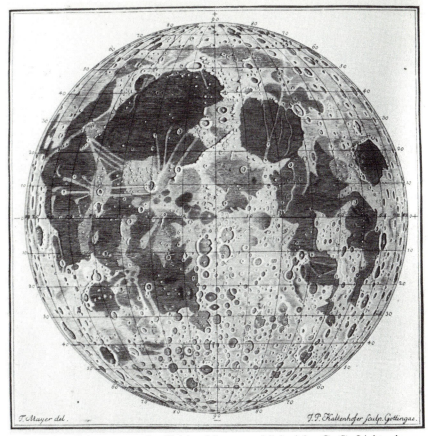

Fig. 14. Small lunar map by Tobias Mayer, published by G. C. Lichtenberg, engraved by J. P. Kaltenhofer

imperishable as the celestial body which it describes.'[32] But the investigations of Schroeter were also soon outdated by still more ambitious ones. W. G. Lohrmann in Dresden planned a comprehensive lunar map which was to have a diameter of 96.5 cm and which would show numerous additional details. In addition to a regular job (like Schroeter), Lohrmann made time to observe, and did so primarily with a Fraunhofer refractor of 120 mm objective size. However, of the 25 sections planned, only 4 were published (1824); not until 1878, over thirty years after Lohrmann's death, did the whole atlas appear through the initiative of J. F. J. Schimdt.

In the meantime Beer and Mädler in Berlin published a lunar map which greatly paralleled the undertaking of Lohrmann. It can hardly be doubted that this map came about because the publication of Lohrmann's sections ceased. The work appeared in 1837 under the title *Mappa Selenographica*. The Moon was reproduced with a diameter of 95 cm. Moreover, a detailed text

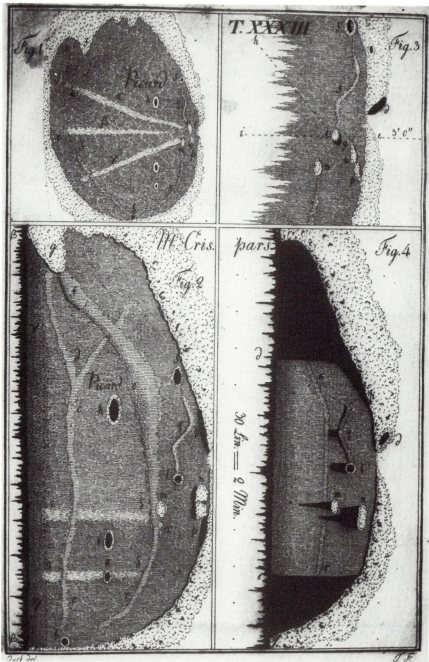

Fig. 15. Detailed maps of the lunar surface by J. H. Schroeter

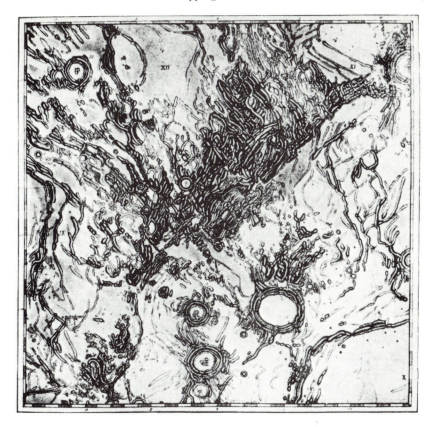

Fig. 16. Section from the lunar map of J. F. J. Schmidt

accompanied the valuable work. It was Bessel's opinion that these maps of Lohrmann, Beer, and Mädler were as good as could be made; in a certain sense he was quite right, although still greater undertakings were carried out. The most important of these maps was made by J. F. J. Schmidt, who worked a total of seven years on these *Charte der Gebirge des Mondes* (Charts of the Mountains of the Moon, 1878). The largest representation of the Earth's satellite and the one richest in detail was made available on a map of diameter 195 cm.

However, Schmidt must have known that it was impossible for an observer to telescopically record all the known details within a reasonable time. Schmidt's lunar map alone contained more than 30 000 craters of all sizes and a number of other details which were lacking in the maps of predecessors (Fig. 16).

The last great German lunar observer, P. Fauth, planned a still larger map. The diameter of the Moon amounted to almost 350 cm (scale

1 : 1 000 000). A complete publication of this map was delayed for many years until 1964.

J. Franz's contour maps of the Moon, made toward the end of the nineteenth century, also deserve mention. He derived the altitude variations of single craters from the average levels and mapped out lines of constant altitude. From this he showed that the mountainous regions were very high, and the *maria* low. As the zero-point level Franz had chosen the average altitudes of the north and south poles of the Moon. Later investigations revised the results of Franz to a certain degree, partly calling them into question. In particular, Hopmann (1952) found that the *maria* on the average do not lie at lower altitudes than the other regions of the lunar surface.

In the meantime photography had also proven advantageous for mapping the Moon. In contrast to drawings, photographs represent an objective record; however, they portray the Moon for fixed Sun elevation angles only, which map-making does not have to do.*

With the two lunar atlases of the Lick and Paris observatories lunar cartography by photographic means reached great heights. The *Lick Observatory Atlas of the Moon* appeared in 1896/97. The total diameter of the Moon amounted to 97.5 cm. The *Atlas photographique de la lune* by M. Loewy and P. Puiseux was published over the years 1896 to 1909.

Concerning the physics of the Moon, topographical investigations only told us that the Moon has no atmosphere. All subsequent details are a result of astrophysics, and include the temperature determinations, the photometric investigations, from which it was attempted to determine the nature of the lunar surface, etc. The Soviet space probes and the manned American flights to the Moon represent a qualitatively higher stage of research concerning our satellite. These projects led to a number of new results and permitted a more detailed inquiry into the nature of this celestial body.

---

* Three works should be mentioned in this discussion of the Moon: Nasmyth and Carpenter's illustrated treatise (1874), Neison's even more elaborate volume (1876), and W. H. Pickering's *An Atlas of the Moon* (a photographic atlas, 1903). (Tr. note)

2

The origin of astrophysics

Because astronomers had such great success with celestial mechanics and positional astronomy, many became blind to other principally new areas of research. Astronomers quite brilliantly did their part to establish a rock-solid foundation for dynamics, and this led them back to philosophical considerations. At the time when the successes of dynamics were achieved, capitalism itself was 'primarily mechanical, for of the natural sciences at that time, only the . . . mechanics of weight had attained an advanced state'.[1] Bessel, the great expert in celestial mechanics and positional astronomy, dogmatically stated in one of his popular treatises that 'astronomy had no other task than to find rules for the motion of every star; its reason for being follows from this'.[2] The observation of planetary surfaces already had as its purpose the investigation of the physical natures of these celestial bodies, but to Bessel this seemed a thankless task. Gauss explained his views concerning the activities of astronomy in a completely similar fashion: 'Speculation in astronomy first ceases and proper knowledge begins with the data which are capable of mathematical expression, such as the size and shape of the celestial bodies, their distances, their corresponding positions and most significantly . . . their motions.'[3] All investigations which were not founded on classical celestial mechanics were either completely avoided or at least regarded with scepticism. For example, in one of his treatises J. H. Mädler posed the question as to whether or not information could ever be 'expected concerning the actual inner nature of the individual fixed stars and their systems', and he came to the conclusion 'that even under the assumption of still greater improvement of optical and mechanical means, no prospect could be expected to make such observations possible'.[4] The founder of positivist philosophy, A. Comte, maintained that the chemical constitution of the Sun must forever remain unknown to mankind; whoever concerned himself with such questions was wasting his time. When the young J. K. F. Zöllner, later a renowned pioneer of astrophysics, expressed to the well-known physicist H. W. Dove the hope of some day being able to learn something of the nature of the stars by investigating starlight, the famed authority replied: 'What the stars are, we do not know and will never know!'[5]

All these were scientific convictions of a number of very successful scientists. As far as historical perspective is concerned, they are only footnotes, for at the time these opinions were published celestial mechanics stood right at the threshold of a new frontier; these pessimistic prognoses were strikingly refuted within a very short time. Characteristically, almost all of the founders of this new branch of astronomy were not astronomers. For the most part they were young physicists who stood completely outside the tradition of classical astronomy.

Similarly, the methodical foundations of the new discipline did not result from astronomy, but from several areas of physics and chemistry which had undergone great advances in the first decades of the nineteenth century, such that their successful application to the study of the stars was made possible. Photometry, spectroscopy, and photography were just such areas. The just-mentioned physicist from Leipzig, Zöllner, a passionate advocate of these new methods for the study of the stars, suggested for this discipline the term generally used today – 'astrophysics'.[6]

Astrophysics did not work its way in without opposition. The number of astrophysical works in the journals of astronomy was extremely small. Even when a wider application of the new methods was used to investigate celestial bodies (*c.* 1895), the proportion of such works in the journal *Astronomische Nachrichten* was only about 6 percent.

Because of the conservative, traditional attitudes of German astronomers, 'astrophysical outsiders' created the foundations of the new discipline. Also, this progressive area of astronomy was already flourishing in the USA at the end of the nineteenth century, where no immutable tradition of classical astronomy had taken root which could hamper its growth.

PHOTOMETRY

The stars were first grouped according to luminosity by Hipparchus (second century BC). A division of stellar luminosities into six so-called magnitudes resulted, containing the brightest stars (1st magnitude) down to the faintest visible with the naked eye (6th magnitude). Luminosity data of this kind are necessarily only crude estimates. This was essentially the state of photometry for over two thousand years. There was also no theory of light measurement. There had to be some specific scientific or practical need in order for the situation to change.

Inspired by Newton's scientific investigations concerning the nature of light, J. H. Lambert published in 1760 the first detailed investigation on the subject of light measurement. The theoretical statements in Lambert's *Photometrie* were epoch-making, especially since the author stated the need for a scientifically-founded theory of light measurement. As Lambert formulated it, we still lacked a measuring instrument for the study of light, analogous to the thermometer for the study of heat. Indeed, Lambert's reproach was hardly the only reason why photometry in the nineteenth century realized such rapid progress. Rather, there were specific interests on the part of the astronomer, and also on the part of technology and industry. Several stars whose luminosities changed became known to astronomers. Why should there not be others? Indeed, the accurate investigation of these objects and the discovery of other variable stars required accurate luminosity measurements. In particular, F. W. Argelander inspired a systematic observation of stellar luminosities with his 'Invitation to friends of astronomy' (1844); he also created an accurate method of estimating magnitudes and expended considerable energy for this new area of research (see pp. 107 ff.).

On the other hand, lighting technology also developed in the nineteenth century. This innovation was of the greatest significance for the development of capitalist production relationships, above all with the implementation of night work and the marked exploitation of manpower and the use of machines. Simultaneously, the production of all kinds of lamps (where success was achieved) was an extraordinarily profitable business which later grew to become a giant industry. By the first half of the nineteenth century London, Paris, and Berlin were lit up by gas lamps. Because it had become practical to 'make night into day', comparisons between the luminosities of the new artificial lamps and the natural light sources, especially the Sun, were an obvious result; here technical and astronomical photometry met. Numerous physicists and astronomers actively concerned themselves with the production of tools for the measurement of light, and many photometers from those years were discussed in textbooks on astronomical photometry as well as in standardized works of lighting technology.

The basic principle of the most important photometers for the measurement of stellar luminosities consisted of the fact that light equality was produced between an object to be measured and an artificial or natural comparison object. The usefulness of the principle is based on the fact that the equality of two luminous sources can be judged by the human eye with incredible accuracy. The way it was done in those old photometers was that one of the two light points, either the star to be measured or the comparison star, was found to vary in its brightness and this variation was measured with the attainment of an equality of luminosity. The dimming was achieved in different ways, by changes of distances, filters, rotating sectors, etc.

The introduction of the visual stellar photometer gave rise to a whole series of theoretical and practical questions, through which the subsequent development of photometry was essentially influenced. Theoretically, it must first be clarified for every type of photometer, how one can derive the

Fig. 17. J. K. F. Zöllner

intensity of the object to be measured from the reading of the photometer, that is, in which way the reduction of the measurements is to be undertaken. A practical demand for the designer of photometers was that a light-sensitive material as large as possible had to produce its effect in the shortest possible time. For the widest application of most of the photometers developed at that time, however, there was no such thing; their availability was far too circumstantial. Likewise, the reduction of the readings was in part quite problematical, for particular scientific facts had to be known; the lack of such facts brought with it great uncertainties. For example, John Herschel compared the luminosity of the star in his photometer with a lunar disc stopped down by a lens. In the reduction process he had to deal with the unsolved problem of the dependence of the lunar brightness upon the lunar phase.

The astrophotometer of Zöllner represented a noteworthy breakthrough in the area of practical stellar photometry. Since his student years in Berlin and later in Basel, the German astrophysicist had concerned himself very intensely with light measurement; even his doctoral dissertation dealt with luminosity measurement and technical light sources. When the Royal Academy of Sciences in Vienna offered a prize in 1857 for the accurate measurement of as many stellar luminosities as possible, Zöllner's teacher suggested that he document all his experience up till then on this subject, resolve some particular questions, and apply for the academic prize himself. So arose Zöllner's classical work which appeared in 1861, *Grundzüge einer allgemeinen Photometrie des Himmels* (Foundations of a General Photometry of the Heavens). Besides many important theoretical considerations, this work contained above all a description of the stellar photometer named after Zöllner, which became a useful model for practical astrophotometry. The measured dimming of an artificial comparison star is obtained with this instrument through polarization and the use of two Nicol prisms. Furthermore, Zöllner's photometer made possible the attainment of color equality between the comparison object and the object for measurement (Fig. 18).

Zöllner only measured the luminosities of 226 stars with this instrument, but his photometer was distributed worldwide. Later, in somewhat modified form, it led to substantial breakthroughs in our understanding of stellar luminosities. Within a few years Zöllner had orders from around the world, and the firm in Gotha which built the instruments supplied 22 in a short time to famous observatories in Russia, America, England, Holland, and other countries.

The introduction of visual measurement in the area of luminosity estimates brought an improvement of accuracy of about a factor of ten. Not until the application of the photoelectric cell in our century has a comparable leap in measurement accuracy been achieved. Today stellar luminosities can be measured to a thousandth of a magnitude.

However, with the appearance of visual measurements of luminosity, the astrophysicists had to address an unsolved problem. This required half a century of lively debate for resolution, in spite of its seeming simplicity. The classical luminosity data in magnitudes, which resulted from the impression of brightness which the eye received, could not directly be obtained with photometers; moreover, these devices measured light intensities, such that the question originated as to what was the actual relationship connecting the subjective impressions and the objective measured intensities. Independently from each other and in different ways, K. A. Steinheil and G. T. Fechner discovered that the light impressions are proportional to the logarithms of the light intensities. This relationship, which on the whole also holds for hearing and taste, is called the Weber–Fechner psychophysical law. However, for astronomers this discovery alone was not yet sufficient. They first had to know whether the relationship was universally

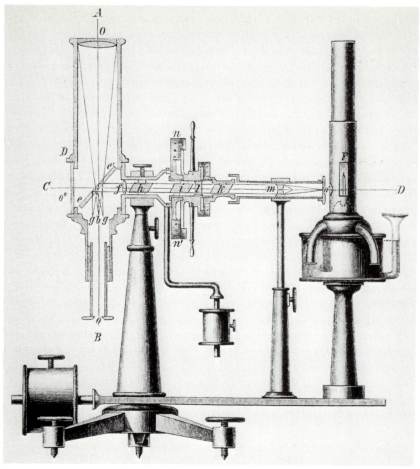

Fig. 18. Schematic diagram of the Zöllner photometer. The light of the star to be measured passes along the telescope axis AB into the ocular *o*; the light of the comparison star is directed to the diagonal mirror along the axis CD and from it also passes into the ocular

valid – whether deviations arose for either large or small luminosities. Secondly, the proportionality constant one had to adopt had to be clarified. Then one had to consider that the older stellar luminosity estimates, which one had to invalidate with these considerations, were extraordinarily uncertain, and as a result were hardly suitable for precise clarification of this difficult question. Therefore, a number of astronomers were keen to throw all of the classical magnitudes out, and instead wanted to use only the logarithms of intensity. Indeed, these ideas do make some sense, but they would make all earlier luminosity catalogs unusable.

Thus a generally accepted relationship gradually developed between stellar luminosity data in magnitudes and the light intensities associated with them. The scale and zero point are chosen such that the classical catalogs retain their usefulness. During the first decades of the twentieth century an international agreement was reached concerning the relationship, whereby the scientific authority of the great luminosity catalogs of the Harvard Observatory and the Astrophysical Observatory of Potsdam were finally considered (see pp. 87 ff.). A suggestion already laid down in 1856 by N. Pogson was adopted for the definition of the scale.*

Besides visual photometry, photographic photometry also progressed very rapidly. The first investigations used to derive stellar luminosities from the dark images of stars on photographic plates were undertaken by G. P. Bond and J. A. Whipple in 1857.

The quite lengthy investigations for the derivation of a relationship between the stellar luminosities and the photographic data such as exposure time and image size simultaneously made a special system of photographic luminosities necessary, for the photographic plate proved to have a different photon response from that of the human eye. Besides this, the system of photographic (blue) luminosities had to be tied to the visual system. Modern astrophysics, in particular the work of K. Schwarzschild, helped us derive the dependence of image size on the photographic plate as a function of exposure time. In order to tie together the system of photographic luminosities and the visual system, in 1910 E. C. Pickering suggested that the stars of spectral type Ao in the magnitude interval from $5^{m}_{.}5$ to $6^{m}_{.}5$ should be considered to have the same luminosity on the average in the two systems. However, the standard system of photographic luminosities (a number of stars in the region of the North Celestial Pole – the polar sequence) exhibited a series of problems owing to difficult practical realizations of this attempted calibration; the polar sequence itself was tied to a standard.

With the steadily increasing knowledge concerning the nature of stellar radiation, in which primary qualities of stars are made known, a serious drawback concerning the previous development of the photometer became evident, namely, that there was no provision for tying together the visual and photographic systems. The development of the two systems now seems to have gone on along completely separate paths. As a result photometry required further physically sensible development so that physical questions could be solved.

Nevertheless, the creation of a classical notion of visual and photographic integral luminosities was not at all in vain, for in the areas of cataloging, stellar statistics, and research on variable stars, classical photometry has quite a considerable significance.

The physical investigations led to the development of a great number of

---

* Each magnitude corresponds to $\sqrt[5]{100} \approx 2.512$ in relative luminosity. For example, a star of magnitude 6 has a luminosity 100 times less intense than a star of magnitude 1. (Tr. note)

photometric systems, which allowed stellar radiation to be measured in a wide variety of ways. The much-used and well-known ultraviolet–blue–visual (UBV) system is the culmination of some of these systems and the prototype of many others.*

SPECTROSCOPY

The dispersion of sunlight by glass prisms was already known by Isaac Newton in 1666. He let white sunlight enter through a round opening in a window shade and then viewed the dispersion behind a prism; he found from this a different refractability of the different rays and noted that white light is made of all the various colors.

However, significant progress was not made concerning the nature of light until Wollaston obtained one important result in 1802. Just as in photometry, the long standstill gave reason for optimism, for scientists had really just begun the investigation of sunlight. However, at the beginning of the nineteenth century the discussion concerning the nature of light once again gathered momentum. Two discoveries are of particular note. When Herschel investigated the distribution of heat in the solar spectrum with a thermometer in connection with his investigations on the construction of telescopes, he determined to his amazement that the temperature maximum lay on the other side of the red light in the invisible. In the next year, 1801, J. W. Ritter made a discovery which was just as interesting. He spread damp, fresh silver chloride on a piece of paper. He tried to observe the change of color of the silver chloride from the influence of the solar spectrum and found that the effect begins near the violet end of the spectrum and there – likewise in the invisible – it was strongest. With this, ultraviolet and infrared rays were discovered. They excited much attention and made spectroscopic experiments once again a high priority item.

Wollaston achieved a substantial result beyond those of Newton, in that he viewed the Sun through a prism and a narrow slit. In contrast to Newton, he observed that specific colors of the solar spectrum do not blend together smoothly but appear interspersed with dark lines among them.

J. Fraunhofer succeeded with a corresponding breakthrough in the years 1812–1814. He had endeavored unsuccessfully for some time to produce perfect achromatic objective lenses for telescopes. For this the indices of

---

* The application of photoelectric photometry to stellar astronomy was first accomplished after about 1912 by a number of independent researchers: P. Guthnick at Berlin, H. O. Rosenberg and F. Meyer in Tübingen, and W. F. Schulz and Jakob Kunz at the University of Illinois. Using photoelectric cells that Kunz built, Joel Stebbins carried out many pioneering investigations at the University of Illinois and the University of Wisconsin, and at Lick and Mt Wilson observatories. The work of Stebbins, Albert E. Whitford, Gerald E. Kron, and Harold L. Johnson in the 1930s, 40s, and 50s paved the way for a great era of photometry. Today, with advances in electronic technology, even amateur astronomers can build photoelectric systems that give measurements accurate to a hundredth of a magnitude or better. (Tr. note)

refraction of glass had to be determined more accurately than was possible with the other methods used at that time. Ideally, one needed to know the refraction of the glass at each specific color. But the colors in the spectrum had no defined boundaries. However, in the spectra of flames Fraunhofer found a sharply defined bright line between the red and yellow colors; this mark was still not sufficient for his purpose. When he later undertook the search for similar lines in the solar spectrum, he saw to his amazement a great number of sharp dark lines, which appeared to be very suitable as standard marks. Therefore Fraunhofer carefully registered them and counted a total of 475 lines. The most prominent lines he designated with capital letters.

The discovery of these Fraunhofer lines (named after their discoverer) was an important step for scientific spectroscopy. They were the practical result desired by Fraunhofer; he could now determine the refractive indices of glass considerably more accurately than before. The careful measurement of the positions of the specific lines led him necessarily to another remarkable fact; Fraunhofer found that the wavelength of the already-mentioned yellow line in the flame spectrum precisely agreed with the value for the dark line in the solar spectrum which he called the D-line.

In the meantime many scholars throughout the world turned their attention to spectra. For example, John Herschel found that flames colored with different chemicals gave different spectra. Similarly, W. H. Fox Talbot proved that one clearly obtains different spectra if one adds lithium instead of strontium to a flame. In the 1820s Talbot suggested the idea that one can eventually learn which substance is contained in the flame from an inspection of the spectrum.

More and more observational material was accumulated which substantiated this idea – a specific connection exists between the spectrum and the chemical composition of the substance which gives off light. All of this, however, was only guesswork and empirical material without a theoretical foundation.

The coincidence of the D-line in the solar spectrum with the yellow line in a flame spectrum correspondingly became important for the real development of spectral analysis and its application to celestial bodies. This fact was generally known to many scholars; however, no one had attempted to give an explanation for it.

Within a very short time the preliminary scientific work of half a century culminated in a general theory as the result of the collaboration of a chemist and a theoretical physicist: R. Bunsen and G. Kirchhoff. One day these two researchers directed their spectroscope at the Sun and simultaneously placed a flame colored with table salt in front of the slit opening; they were of the opinion that the dark D-lines in the solar spectrum would shine brighter in this way according to the 'summation of light'. However, to their great perplexity precisely the opposite happened; the lines appeared even

Fig. 19. G. Kirchhoff

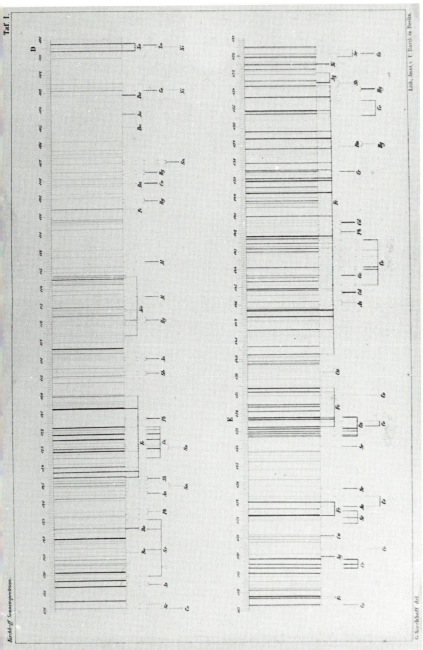

Fig. 20. Portion of the solar spectrum according to G. Kirchhoff

darker! Kirchhoff commented on this singular result with the remark: 'This is either nonsense or a very remarkable thing.'[7] The next day he suggested the following explanation of the notable phenomenon: the sodium vapors absorb rays of the same color which they emit in the glowing condition. Subsequent experiments immediately showed that this was true.

Now it seemed quite likely that the many other Fraunhofer lines came about in the same way as the D-lines. Kirchhoff next was able to confirm this hypothesis when he found a total of 70 bright lines of iron vapor corresponding to 70 dark lines in the solar spectrum. The existence of a marked relationship between light emission and light absorption clearly manifested itself with this. This assumption led Kirchhoff to the well-known rule of thumb: the ratio of the coefficients of emission and of absorption is a function of wavelength and temperature, and is the same function for all bodies. With this the preliminary history of quantum theory begins; Kirchhoff's rule proved to be the basic idea for the explanation of spectra, and with it we had a starting point for the understanding of the processes going on in the stars (see pp. 119 ff.). With the determination by Kirchhoff and Bunsen that every chemical element has a corresponding spectrum, spectral analysis was born!

Scientists tried to determine how the presence of the most insignificant traces of elements could be demonstrated; at the same time this pushed our horizons out into cosmic space and laid the groundwork for the investigation of the gaseous composition of celestial bodies. *Chemische Analyse durch Spectralbeobachtungen* (Chemical Analysis by Spectral Observations – the classical work by Kirchhoff and Bunsen from the year 1860) was a genuine scientific sensation and a turning point for all those who had been of the opinion that there were no practical possibilities of discovering something concerning the physical and chemical nature of celestial bodies.

During the 1860s spectral analysis experienced a flourishing without equal. The number of publications rose precipitously. Above all, the particular developments showed that many other scientists were very interested in a rapid development of the new methods of investigation. The chemists discovered ten new elements within a relatively short time through the application of spectral analysis, among them helium in the Sun (1868). Above all physicists saw in spectral investigations the possibility of obtaining exact wavelengths. The demonstrated spectral patterns were important for the empirical foundation of atomic theory at the beginning of the twentieth century (see pp. 115 ff.). But spectral analysis revolutionized astronomy – light rays carry information concerning the chemical composition of the celestial bodies over millions and millions of kilometers, enabling us to study these bodies.

As a result the investigation of the chemical composition was only a beginning. Over the course of time consequent development of spectral analysis led, in close connection with physics, above all to more and more penetrating insights. For example, in 1880 G. Wiedemann determined the influence of temperature, pressure, and density of the absorbing layer on the darkening and profile of the Fraunhofer lines. Later it was shown that we can investigate magnetic and electric fields from the splitting of spectral lines (Zeeman and Stark effects).

PHOTOGRAPHY

The invention of photography by L. J. M. Daguerre and J. N. Niepce came about in 1838 and was the result of a long sequence of earlier investigations by different chemists, who concerned themselves with chemical reactions triggered by light. The famous French astronomer F. Arago, one of the first initiated by Daguerre into the secrets of the new process, presented a provisional, detailed report about it to the Paris Academy of Sciences in January 1839. With scientific foresight and unmistakable good sense for the capability of development of the sensational novelty, Arago already sketched out in this treatise a picture of the eventual applications of the new process. Astronomy provided certain opportunities too, for Daguerre himself already had undertaken an investigation, on the basis of a suggestion of Alexander von Humboldt, to photograph the Moon – actually with very little success; the Moon left only a white smear of an image, nothing more. In spite of this, however, the astronomers throughout the world kept an ear open, for they were interested in anything to do with light. Thus it was in 1839 that two astronomers suggested the name still used today for this process: J. H. Mädler and John Herschel called Daguerre's invention 'photography'.

While interest in photometric and spectroscopic questions in the scientific and technical worlds was just around the corner, the enthusiasm for photography was felt by practically everyone. The possibility of building a lucrative business from the photographic reproduction of buildings and people immediately resulted from this. Factories developed optical instruments and photographic apparatus, and all sorts of photographic studios came into being. Outstanding daguerrotypes excited the greatest interest at the world's fairs in the middle of the century. As a result the development of the process made very rapid progress, which allowed interest in all circles to grow on its own, especially, however, in art and science. Even with astronomical photographs businesses were possible, as the sale of the best lunar photographs obtained by the American pioneer of photography L. M. Rutherfurd proves. W. H. Fox Talbot's invention of the paper negative and the whole negative–positive process gave us the first means of reproducing photographic pictures. Finally, the invention of the dry silver bromide plates in the 1860s brought about a specialized photographic industry. Photographic technical journals and associations sprang up like mushrooms in America, England, greater Germany, and in France. Two of the

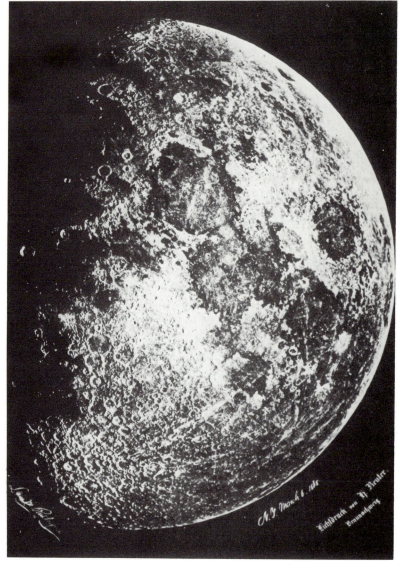

Fig. 21. Lunar photograph by Rutherfurd from the year 1865

best known pioneers of astrophotography – H. Draper and L. M. Rutherfurd – belonged to the American Photographical Society, founded in 1859.

Draper's father had photographed the Moon as early as 1840. The son's
earliest lunar photograph, still preserved today, was made in the year 1863
and was taken with a 40-cm reflector. Rutherfurd obtained substantially
sharper and more detailed lunar photographs in the 1860s (Fig. 21).

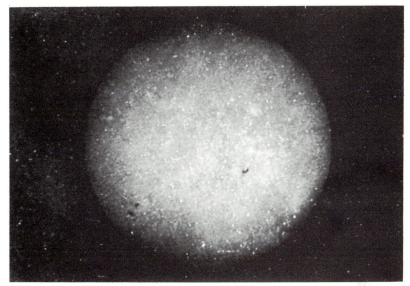

Fig. 22. Daguerreotype of the Sun from the year 1845

The first photographic renderings of the Sun left something to be desired, for here one had to solve the technical problem of very short exposure times. In the other photographic experiments this question did not come up, for the exposure times in general amounted to several minutes. After about 1845 solar photography had also made progress; John Herschel suggested that the method of photography should be used to continuously monitor the Sun, particularly to make the sunspots on the pictures more evident.

However, relatively rapid successes were not achieved in spite of these optimistic feelings, and photography did not work its way into astronomy without resistance. There were instead many sceptics whose arguments were based on the defective quality of the pictures. The level of detail in the Moon's surface rendered at that time in drawings could in no way be matched by photographs. Photographing stars was even more difficult. Although under the direction of the astronomer W. C. Bond, the photographer J. A. Whipple at Harvard Observatory succeeded in photographing the brightest star in the northern sky, Vega in Lyra; this remained for the next few years the only success. The longer exposure times required convincingly demonstrated the necessity of having automatic clock drives for the equatorially-mounted telescopes. In 1824 Fraunhofer outfitted the refractor at Dorpat with such a device for the special purpose of double star observations, but in general this technique was not widespread.

The technical possibility of even longer exposure times and the steady increase of sensitivity of the photographic material by means of uninterrupted developmental work soon permitted the photography of very faint

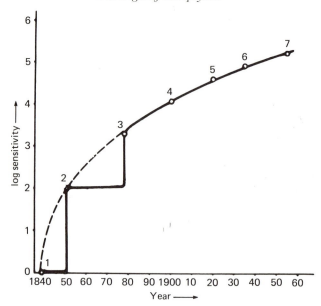

Fig. 23. The increase of light sensitivity of photographic plates

astronomical objects (Fig. 23). Thus in 1865 Rutherfurd photographed stars down to a brightness of $9^{m}.5$ in 30 minutes with a telescope of $27^{m}.5$ cm objective size.

The subsequent use of photography involved the production of special refractors, which later developed into the astrograph, an instrument exclusively used for celestial photography.

Stellar photography then showed it to be possible to accomplish many different measurements in the laboratory which until then had to be undertaken under the most unfavorable conditions right at the astronomical observational instrument.

In 1857 Bond had undertaken distance measurements from a series of photographs of double stars, which pointed out a small error in the famous micrometer measurements of Wilhelm Struve. In spite of these hopeful beginnings, photography could not yet be comprehensively applied to positional astronomy. The astronomers gained some experience in 1874 on the occasion of the Venus transit of that year. They tried to obtain the required data with the help of photography, but the results were so unsatisfactory that photography was not used to record the second Venus transit of the nineteenth century in 1882.

Unequivocally, celestial photography proved to be much more effective for the investigation of nebulous objects than visual telescopic observations.* Already in the 1880s the pictures of the Orion Nebula exhibited more details than could be discerned by the eye itself and the best telescope. Above all the subjective representation of objects gradually was phased out in favor of the new objective process (Figs. 24a and b).

When Max Wolf and E. E. Barnard used short-focus objectives for celestial photography, nebulous forms appeared on the plates which no eye had ever seen before in the sky. Unexpected secrets were elicited from the cosmos through the use of photography.

According to all scientific expectations which one associates with photography, Arago indeed proved to be right when it was shown that one could essentially measure what one could not previously see. The rapid development of astrophotography progressed so impressively that the Hungarian astrophysicist N. Konkoly even published a monograph in 1887 – *A Practical Guide to Astrophotography*.

All astrophysical processes developed at first together, for they all overlapped. Thus photometry exhibited significant progress due to its connection with photography after the foundations for the development of photographic photometry were laid down according to Schwarzschild's explanation of the quantitative interrelationships between light intensity, exposure time, and image density.

Spectrography arose from the connection of spectroscopy with photography. Spectroscopy would have remained in a very primitive condition without the photographic recording of the spectrum and the possibility of specially designed laboratories.

Also of greatest significance was the origin of spectrophotometry from the combination of spectroscopy and spectrography with photometry. Fraunhofer had already studied the distribution of brightness of the Sun's spectrum. He used for this a visual photometer, with which he compared the brightness of a standard light source with the brightnesses of different parts of the solar spectrum. For reference he chose the intensity of the spectrum between the D and E lines. Although his results were sketchy (for example, he did not know the spectral sensitivity function of the human eye), these investigations historically represent the first spectrophotometer. Above all, the enormous significance of spectrophotometry at the beginning of the twentieth century is explained by the well-known fact that important physical information is contained in the brightness distribution of spectra (see pp. 121 ff.).

---

* The brothers Paul and Prosper Henry took many splendid pictures of the planets, star clusters, and nebulae with a 13-inch refractor at the Paris Observatory. A three-hour exposure of the Pleiades taken by them in 1885 registered over 1400 stars and clearly showed the faint nebulosity only suspected by naked-eye observations. Indeed, their photographic techniques and equipment were so well regarded that the international program for photographing the sky, organized at Paris in 1887 (see p. 33), called for the participants to use refractors identical to that of the brothers Henry. (Tr. note)

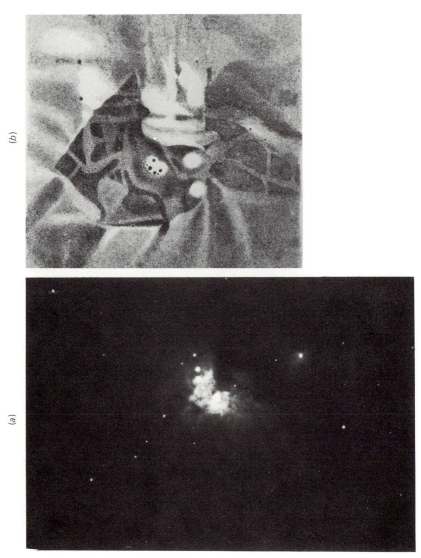

(b)

(a)

Fig. 24. (a) The Orion Nebula, photographed with the large telescope of the Astrophysical Observatory of Potsdam on 1 November 1901; (b) The Orion Nebula

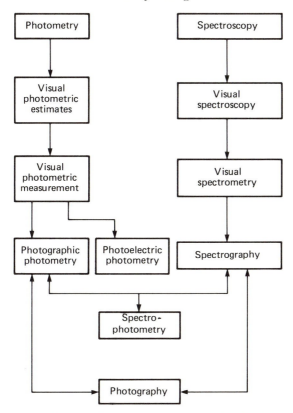

Fig. 25. The development of astrophysical methods

However, astrophysics in the nineteenth century did not limit itself to the elaboration of fundamental research methods and development of equipment. Rather, the first attempts at solving special questions were simultaneously undertaken and large-scale data bases were compiled along the way.

Thus even in the nineteenth century arose great luminosity and spectral surveys of the sky (pp. 87 ff.), new areas of research like solar physics (pp. 98 ff.) and the broad field of variable stars (pp. 107 ff.); new insights into the nature of comets and nebulae were achieved (pp. 112 ff.), and, with the application of the Doppler principle, velocity components of astronomical objects were measured for the first time (pp. 94 ff.).

LUMINOSITY CATALOGS

The invention of photometric measuring devices allowed much more accurate luminosity measurements to be made. However, they still were not of great specific scientific value. For example, Zöllner explained in 1881:

'Although it will one day be possible to obtain the observational material gathered with photometric means in a similar manner . . . for furthering our knowledge concerning the cosmos, as this did happen in such a brilliant and successful way with positional determinations – this will have to be put off for some time.'[8] In spite of the unresolved question of the use of photometric investigations the astronomers nevertheless dedicated themselves with great pageantry to this problem and within a short time completed comprehensive sky surveys; several years later Zöllner formulated a series of questions concerning the nature of the stars which could be clarified with the help of photometry.

The most comprehensive compilation of photometric data is contained in the large-scale survey of the sky carried out by Argelander and his colleagues E. Schönfeld and A. Krüger at the Bonn Observatory from 1852 to 1868 (the *Bonner Durchmusterung**: *BD*). With tremendous perseverence the authors of this famous work determined the approximate positions and estimated the luminosities of all stars to a limiting magnitude of $9^m.5$ from the North Celestial Pole to a declination of $-2°$. The Fraunhofer comet finder with a diameter of 76 mm (focal length 650 mm) served as the observational instrument. The diameter of the field of view amounted to 6°. Almost a million luminosity estimates went into the making of the catalog.

Three periods are to be distinguished in the compilation of the catalog. Until 1854 the estimates were made to the nearest whole magnitude (20% of all observations). In the second period (till 1857) half-magnitude estimates were introduced (50% of all observations). The remaining objects were estimated to the nearest 0.1 magnitude. Though the luminosity information was not the main part of the work, these data were not consistently obtained under good seeing conditions. Schönfeld's continuation of the *BD* for the southern sky (133 659 stars) should also be mentioned. Then J. Thome, R. H. Tucker, and C. D. Perrine made observations for the *BD* for stars even further south. From 1885 to 1908 they observed more than half a million stars to a limiting magnitude of 10 and to declination $-62°$. The average uncertainty in the luminosity estimates was $\pm 0^m.2$.

Because of their scope, these works alone are historically valuable. Even later, when more accurate luminosity estimates were available, reference was made again and again to these surveys because it was clear that the astronomers had carefully considered the relations between the luminosity systems given in these catalogs and the other observations.

Improved observing methods very soon led to substantially more accurate results. With the use of photometric devices new surveys were started. These were not quite so comprehensive as the just-mentioned work. The most important investigations of this kind were begun by E. C. Pickering

---

* Literally, *Bonn Survey*; this work is always referred to by its German title, so we shall leave the German word for survey as it stands. (Tr. note)

and C. Pritchard, as well as G. Müller and P. Kempf at the observatories in Cambridge (USA), Oxford, and Potsdam.

Measurements made at Harvard College Observatory under Pickering's direction were carried out from 1879 until the publication of the *Harvard Revised Photometry* in 1907. Pickering and his co-workers used a polarizing meridian photometer with a Nicol prism. The Pole Star temporarily served as the comparison star; then other stars near the pole were used. Of great practical significance was the fact that E. C. Pickering carried out his photometric work with the notion of magnitudes in mind and tied it all together with the scale already suggested in 1856 by N. Pogson.

The Oxford photometry was carried out with the use of a Keil photometer and was of interest as far as methodology goes. It was essentially a repetition of the Harvard photometry.

The Germans Müller and Kempf derived very accurate stellar luminosities in the *Potsdamer Durchmusterung*, still greatly used today. From 1886 to 1905 the luminosities of all *BD* stars down to a magnitude of $7^{\mathrm{m}}.5$ were measured at the Astrophysical Observatory of Potsdam. The stimulus for this survey goes back to Zöllner, who had proposed the project to his student H. C. Vogel. Later, in his capacity as director of the Potsdam Observatory, Vogel helped realize the project. For the photometry Müller and Kempf used a specially modified Zöllner photometer.

The first portion of the *Potsdamer Durchmusterung* contained the 3522 objects of the declination zone 0° to 20°. The two authors used about 45 000 individual measurements for the luminosity determinations of these stars. The observations for this portion alone of the whole work required 405 nights (during the years 1886–1893). At the same time Müller and Kempf very carefully measured the luminosities of 144 fundamental stars; these were used as the standard stars for the calibration of the other measurements.

The whole work was finished in 1907, but precise results were never achieved with visual photometric procedures. Like the Harvard photometry, the Potsdam system was also connected to the *BD* and used the scale suggested by Pogson.

When the *Potsdamer Durchmusterung* appeared the astrophysicists were already furiously working on the problem of stellar evolution (see pp. 123 ff.). The luminosity measurements were considered very important for the eventual solution of such questions – especially Zöllner's theory of stellar evolution. It was thought that the luminosity variations of the stars (which were only made apparent after great amounts of time) could also tell us something of the variations of the physical nature of the stars; on the other hand the luminosity measurements could 'offer a certain alternative to the great difficulties associated with parallax measurements', such that it also appeared in this regard as the duty of the age, 'to trace out a true picture of the stellar heavens with all existing means at one's command and with this to hand down to subsequent generations a reliable foundation for further speculation concerning processes going on in the universe'.[9]

The introduction of photographic photometry represented a considerable improvement in efficiency in this type of work; the toilsome work at the telescope was eliminated and could be replaced by work in a laboratory. At the telescope the plates were exposed only, and numerous objects could simultaneously be recorded.

The actinometry of K. Schwarzschild at Göttingen (1910) is one of the most significant photographic stellar luminosity surveys. Schwarzschild hoped to measure the luminosities of all stars brighter than 7^m5 in the declination zone from 0° to 20°. Originally conceived as a photographic companion to the *Potsdamer Durchmusterung*, the undertaking was essentially not completed, though the results were more accurate than those of the *Potsdamer Durchmusterung*.

The observatories at Pulkovo, Leiden, and other places published very extensive lists of photographic luminosities.

J. Kapteyn's Plan of Selected Areas is also to be counted among the luminosity surveys. This project's purpose was the determination of all obtainable stellar data, among them also luminosities, in different areas of the sky. These investigations followed the specific goal of investigating the structure of the Milky Way (see pp. 136 ff.). Observatories throughout the world took part in the work for these data compilations, among them the observatories in Groningen, Göttingen, Potsdam, and Cambridge (USA).

Table 2. *Famous luminosity catalogs of the nineteenth and twentieth centuries*

Catalog and compiler	Year of publication	Number of stars	Declination range
Uranometria Nova (Argelander)	1843	3 256	+90° to −35°
Bonner Durchmusterung (Argelander)	1859–1862	324 198	+90° to −2°
Uranometria Argentina (Gould)	1879	7 756	+10° to −90°
Uranometria Oxoniensis (Pritchard)	1885	2 784	+90° to −10°
Southern Bonner Durchmusterung (Schönfeld)	1886	133 659	−2° to −22°
Cordoba Durchmusterung (Thome)	1892–1914	578 802	−22° to −62°
Harvard Revised Photometry (Pickering)	1907	9 110	+90° to −90°
Potsdamer Durchmusterung (Müller and Kempf)	1907	14 199	+90° to 0°
Göttingen Actinometry (Schwarzschild)	1910	3 689	+20° to 0°
Selected Areas, Harvard–Groningen–Durchmusterung (Pickering, Kapteyn, and van Rhijn)	1918–1924	250 000	+20° to 0°

SPECTRAL CLASSIFICATION AND SPECTRAL CATALOGS

J. Fraunhofer aimed his spectroscope at the stars in 1817 and found that various stars had greatly different spectra. Later L. M. Rutherfurd and the Englishman William Huggins confirmed this. Huggins also pointed out that there are distinct spectral types for stars. In 1866 the Italian astrophysicist A. Secchi divided the spectra into three classes and with this put together the first spectral classification scheme. He differentiated the spectra of white, yellow, and red stars (see Table 3 and Fig. 26).

Secchi soon perceived that the great variety of stellar spectra could not be adequately represented by this strict classification scheme; rather, that intermediate examples were to be found which he explained in terms of differences in the stellar atmospheres.

The classification criteria were understandably still quite uncertain. When Secchi had investigated a greater number of stellar spectra over the course of years, he enlarged the scheme to include other classes (1868, 1878).

In 1874 H. C. Vogel also created a classification system. However, he did not organize it, as did Secchi, solely from the subjective appearance of the spectra, but proceeded according to the notion that a classification system must reflect the stages of development of the celestial bodies. For this he considered the notions of stellar evolution in vogue at that time and built them into the classification scheme for stellar spectra. But our knowledge of stellar evolution at that time was quite limited. The English astrophysicist J. N. Lockyer proceeded along similar lines after about 1888. However, the attempt to organize the classification system according to the ideas of stellar evolution did not work out because changes in the theories necessitated changes in the spectral classifications. E. C. Pickering, Williamina Fleming, Antonia C. Maury, and Annie J. Cannon of Harvard Observatory therefore based their classification scheme on the preliminary work of Secchi. In addition to this, substantially better equipment was available in 1890, such that their stellar spectra exhibited many more details. Therefore, Pickering and Miss Fleming also introduced a significantly greater number of stellar

Table 3. *The spectral classification of Secchi*

Type	Spectral characteristics	Examples
I	Hydrogen lines and several lines in the yellow and green part of the spectrum on the continuous background	Sirius, Vega, Rigel
II	A great number of lines in all regions, similar to the spectrum of the Sun	Arcturus
III	In addition to blended sets of dark lines, tinted bands appear	Antares, Betelgeuse

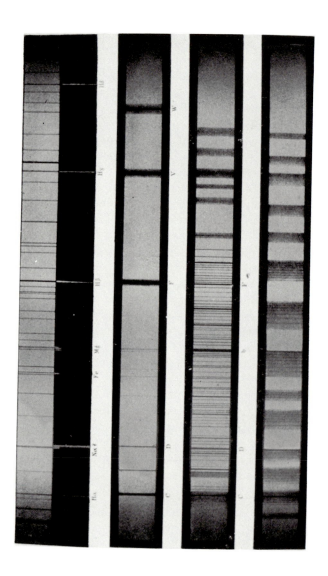

Fig. 26. The spectral types of the stars according to A. Secchi

Table 4. *Comparison of different stellar classifications*[10]

Secchi	Vogel	Maury	Cannon	Examples
(V)	II b	XXII	Oa, Ob, Oc, Od, Oe	Wolf–Rayet stars; γ Velorum, Harvard Annals 50 No 2583, No. 6249, λ Cephei, 29 CMa
		I	Oe 5	λ Orionis
	I b	II	B0	ϵ Orionis
		III	B 1	β Cephei
		IV	B 2	γ Orionis
		IV–V	B 3	π^4 Orionis
		V	B 5	η Tauri
		VI	B 8	β Persei
		VI–VII	B 8, B 9	α Delphini
I	I a	VII	A 0	α Lyrae
		VIII	A 0	α Geminorum
		IX	A 2	δ Ursae majoris
		X	A 5	α Aquilae
		XI	F 0	γ Bootis
		XI–XII	F 2	ζ Leonis
		XII	F 5	α Canis minoris
		XII–XIII	F 5	α Persei
		XIII	F 8	α Ursae minoris
II	II a	XIV	G 0	α Aurigae, Sun
		XIV–XV	G 5	β Bootis
		XV	K 0	α Bootis, α Cassiopeiae
		XV–XVI	K 2	γ Aquilae
		XVI	K 5	α Tauri
III	III a	XVII	Ma	β Andromedae
		XVIII	Ma, Mb	α Orionis, β Pegasi
		XIX	Mb	ρ Persei
		XIX–XX	Mb, Mc	α Herculis, R Lyrae
		XX	Md	o Ceti
IV	III b	XXI	N	19 Piscium
(VI)			R	BD $-10°$ 5057, $-3°$ 1685

Table 5. *Famous spectral catalogs of the nineteenth and twentieth centuries*

Name of spectral catalog or name of author	Year of publication	Number of objects
Suggli Spettri Prismatici dei Corpi Celesti (Secchi)	1867	316
Fleming	1890	10 351
Vogel and Müller	1893	4 051
Lockyer	1902	470
Cannon	1909	1 122
Henry Draper Catalogue (Pickering, Cannon)	1918–1924	225 300
Selected Areas (Becker, Payne, Schwassmann)	1929–1931	about 35 000

spectral types into the first phases of their work, which they designated with the capital letters of the alphabet B to Q. In 1901 Miss Cannon then added decimal subgroups and rearranged the sequence such that the classes O and B came before A. With this the spectral classification of stellar spectra generally used today was fundamentally completed (see Table 4). Later supplements and refinements were the result of new findings obtained with new equipment, for example the introduction of the classes R and S (1908 and 1922).

The International Solar Union already had adopted the Harvard classification system in 1913. With a number of additions the International Astronomical Union recommended it for general use in 1922.

The classification of spectra into different classes was carried out according to certain points of view; one did not want to ignore the fact that star colors ranged from blue to red, the so-called cooling sequence. The Harvard classification therefore became a temperature scale; the spectral classes proceeded from high to low temperatures. The international adoption of the system depended greatly on the fact that the appearance of spectra is determined in essence by the temperature of the stellar atmospheres. In 1921 the Indian astrophysicist M. Saha and the Englishman R. H. Fowler determined the physical reasons for this. Another reason for international acknowledgment of the Harvard system was the fact that this system was accessible to all astronomers in a usable form; the Harvard collaborators published their results in a large spectral survey of the sky, the *Henry Draper Catalogue*. It appeared in the years 1918–1924 and supplements were later published. Other surveys had already preceded this great undertaking (Table 5).

RADIAL VELOCITIES

In 1842 the Austrian physicist C. Doppler showed on theoretical grounds that the frequency which one measures for a source of light or sound will change if the distance between the source and observer is decreasing or increasing. The validity of the Doppler theory can easily be demonstrated for sound waves by listening to an approaching train whose whistle will change pitch as the train passes the observer and then recedes into the distance. On the other hand, it was hard to prove this for light waves because the changes in frequency are easily measurable only if the radial velocity of the light source is a sizeable fraction of the velocity of light. Doppler maintained that the stars manifested such great velocities; the different stellar colors, he contended, were the result of the different velocities, while the natural color of all stars was white. Understandably, this supposition elicited considerable controversy, especially since the dependence of the color of glowing substances upon the chemical composition and other factors was already known. In addition to this, the exaggerated polemical way with which Doppler defended his interpretation of stellar colors proved to be a serious hindrance to the acceptance of his theory.

A new situation did not arise until the creation of spectral analysis. At that time the young German physicist Ernst Mach studied this problem and proved that important information concerning radial velocities is contained in the light of the stars. The changes of color resulting from the motion must be so small, however, that proof was not possible with the naked eye. In order to prove the existence of minute frequency shifts Mach used spectroscopic methods, as suggested by Kirchhoff. The radial velocity can be measured as a shift of the lines with respect to the lines of a laboratory spectrum; if the distance to the object is increasing the lines are shifted toward longer wavelengths (red shifts), and if the distance is decreasing, the lines are shifted toward shorter wavelengths (blue shifts).

After this rational foundation of the Doppler theory was laid out, it was clear how one could confirm the effect of velocity on wavelength. Indeed, success in this area only came to those who were convinced of the demonstrability of the line shifts, for only these scholars set about to discover such shifts.

One of the most vocal advocates of such research was Zöllner. He developed specifically for this purpose the so-called reversion spectroscope, with which the line shifts could be enhanced, and suggested to his student H. C. Vogel that he measure with this instrument the line shifts in the spectrum of the limbs of the Sun, which are to be expected as a result of the solar rotation. This was achieved in 1871, but without the desired degree of accuracy.

Five years later when C. A. Young used a Rutherfurd diffraction grating he could measure the shifts substantially more accurately. His value of the rotational velocity of the Sun derived in this way was in agreement with the value based on the movement of sunspots.

W. Huggins started measuring the radial velocities of stars in 1867. Vogel likewise began his own observations. However, the visual investigation of the stellar spectra and the derivation of line shifts exhibited great difficulties. The results were very uncertain, and it was suggested that further observation should not be made.

Yet Vogel returned to this problem later. His principal sustaining thought was to use photography, a method which had been greatly improved in the meantime. In April 1887 Vogel had attended the International Congress for Astrophotography in Paris and was presumably encouraged by what he learned. Scarcely a year later he submitted to the Royal Academy of Sciences in Berlin the first results of his research under the title 'Concerning the determination of the motion of the stars in the radius of vision through spectrographic observation'. In it he and J. Scheiner published the Doppler shifts of lines in the spectra of the stars Sirius, Procyon, Rigel, and Arcturus. The results excited the strongest international attention. With this the era of stellar radial velocity measurements began. The accuracy was ten times better than with visual measurements. The radial velocities of stars could now be determined to within a few km/s.

The discovery of spectroscopic double stars and the accurate investigation of the Algol system (see pp. 110 ff.) are to be counted among the most convincing successes which Vogel achieved with his collaborators. The significance of the Doppler principle for astronomical research was clearly demonstrated. Numerous foreign astronomers came to Potsdam to learn the technique of measuring stellar radial velocities from Vogel and his collaborators, among them E. B. Frost of Yerkes Observatory and A. A. Belopol'skiĭ, the well-known Russian astrophysicist of Pulkovo Observatory.

In 1877 W. Abney demonstrated that one could determine the rotational velocity of stars from the line broadening in the spectra. It is interesting to see the arguments with which Vogel – the pioneer of spectrophotography – criticized Abney's optimism: 'How Mr Abney . . . can speak of the hope that photography of stellar spectra could be of importance for the question suggested by him is completely incomprehensible to me, for with such faint images as stellar spectra are in general, photography will always be inferior to the usual ocular demonstration.'[11] Actually, photography was employed in this area with results just as revolutionary as in the area of classical determinations of radial velocities. The extensive investigations of stellar rotational velocities, especially the dependence of the value obtained on the stellar spectral type, yielded important, interesting results for cosmogony. Pickering felt that the radial velocities of many stars could be measured from objective prism plates containing the spectra of many objects taken simultaneously. His first attempts relating to this date from the year 1891. Unfortunately, one cannot copy a comparison spectrum onto the plates. In order to bypass this difficulty Pickering tried to place an artificial absorption line on the plates, for which he placed a mask with a special opening in front of the photographic plate. Schwarzschild, who used this method, obtained

Fig. 27. H. C. Vogel

quite accurate, usable results (1913). Still more suitable was the Pickering 'reversion method', whereby the same stellar region is photographed twice on the plates with an objective prism; for the second photograph the prism is rotated 180 degrees around the camera axis. The two spectra for each star running opposite each other then show the doubled shift of the lines. This method was also used by Schwarzschild and was theoretically worked out (1913).

SOLAR ASTROPHYSICS

Kirchhoff's discoveries in the area of spectral analysis paved the way for a
scientific study of the Sun. For the first time it was possible to describe the
physical nature of the Sun on the basis of scientifically gathered data.
Kirchhoff himself concluded that the solar core was a source of extremely
intense white light, and consequently the source of a continuous spectrum.
The observation of the dark lines necessarily implied the existence of a
cooler solar atmosphere, because absorption lines are caused by a source of
continuous radiation and a volume of cooler gas in between this source and
the observer. This is how it was found that the Sun has a greatly heated core
and a gaseous atmosphere of lower temperature.

Kirchhoff believed that the core of the Sun must be solid or liquid, for he
could only explain the continuous spectrum in this way; it was not known
until later that extended layers of gas, heated and under high pressure, emit
a continuous spectrum.

Secchi was the first to advocate the notion that the solar core is gaseous
and that the solar temperature steadily decreases from the core to the
photosphere. However, he could not categorically prove this because he had
no exact information concerning the temperature of the solar surface at that
time. Therefore, the suggestion of relating the amount of energy received by
the Earth to the energy radiated by the Sun was important. The measurement of the amount of energy falling perpendicular on the Earth per unit
area and time (the solar constant) was no simple technical problem;
however, still more difficult was the determination of the effect of the
absorption in the terrestrial atmosphere on sunlight. But to derive the
temperature of the solar surface from the solar constant one needs to know
the relationship between luminosity and temperature. This relationship,
now known as the Stefan–Boltzmann equation, was not discovered until
1879. Until then two other relationships were used, one by Newton and
another by Dulong and Petit. The Newtonian law said that the luminosity
and the temperature were proportional; this idea was generally accepted by
physicists at the beginning of the nineteenth century. In 1817 Dulong and
Petit derived another equation from a series of experiments; they found that
if one considers an arithmetic increase of temperature, the emitted radiation
increases geometrically. Both equations turned out to be wrong and led to
greatly different values for the Sun's photospheric temperature. For
example, C.-S.-M. Pouillet derived a solar temperature of 1750°C with
Dulong and Petit's law, while in 1860 J. J. Waterston obtained a value of 13
million degrees for the solar temperature using the Newtonian Law. Not
until the end of the century did the Stefan–Boltzmann law give us to the real
temperature in connection with the very precise values of the solar constant
obtained in the meantime, such as the result of W. E. Wilson (1901), which
led to a value of the solar temperature of 6590°C.

With the creation of spectral analysis it became possible to investigate the composition of the Sun, if only one could measure as many lines as possible in the solar spectrum. Kirchhoff and his student K. Hofmann made the first attempt at this and published the first atlas of the solar spectrum in 1860/61. However, the wavelengths listed in this atlas were not precise because the scale varied with wavelength due to the prismatic dispersion, and the wavelengths themselves were given in arbitrary units. As a result A. J. Ångström published in 1869 a new atlas of the solar spectrum, which was made by means of a grating and used for the wavelength data a special unit based on the metric system, now known as the Ångström (10^{-10}m). The photographic atlas of the solar spectrum, which H. A. Rowland published in the 1890s and which is still used today, represented a qualitatively more substantial achievement. It covered the wavelength region 3000 to 7000 Ångströms and contained about 20 000 lines.

On the basis of the corresponding absorption lines Kirchhoff had already succeeded in demonstrating the existence of a whole series of elements in the solar atmosphere, among them calcium, copper, barium, strontium, magnesium, nickel, cobalt, iron, zinc, and gold.

The ever more detailed study of the solar spectrum led in 1863 to the discovery of lines which resulted from the absorption of sunlight in the Earth's atmosphere. Specifically, Brewster and Gladstone found that certain lines in the solar spectrum only appear when the Sun is low in the sky and that others become most evident when the Sun is high in the sky. Large-scale investigations in the laboratory by Janssen showed that this was primarily due to absorption by atmospheric water vapour.

Important information was expected from the direct spectroscopic observation of the solar limb regions. This becomes possible during a total solar eclipse. After 1860 the total solar eclipses therefore were of great interest to all astrophysicists. Never before were so many expeditions planned to regions where totality was visible as during the infancy and childhood of astrophysics.

The first results were obtained during the unusually long eclipse of 18 August 1868 (nearly $6\frac{1}{2}$ minutes of totality). Although the spectroscopes used for the investigation of the solar limb regions were still quite crude, the disappearance of the well-known absorption lines was directly discovered. A weakly-distinguished yellow emission line was then interpreted as an indication of sodium. But then the French astrophysicist P. J. C. Janssen proved that the line had a different wavelength from that of the sodium line, and J. N. Lockyer came to the conclusion that this line could not be reconciled with any known terrestrial element. This is how the element helium (from *helios* – the Greek word for Sun) was discovered. Proof of its terrestrial existence was not obtained until 1895 by W. Ramsay.

During the solar eclipse of 1869 C. A. Young and W. Harkness discovered another line which could not be shown to have been caused by an element

Fig. 28. The solar corona during the total solar eclipse of 18 July 1860, photographed
by A. Secchi with a 40-second exposure time

known on the Earth. They thought it was another typical 'solar element',
and it was given the name 'coronium'. Later it was shown that this
interpretation was incorrect.*

Also of great interest was the detailed investigation of prominences,
which were first discovered in 1842 during a solar eclipse. Limited understanding at best was achieved concerning the nature of these formations.
Some people considered them optical illusions; others considered them to
be phenomena related to the lunar limb. Spectroscopic investigations
showed that they are gaseous forms primarily made of hydrogen. The
contradictory reports of various observers made it desirable to investigate
these phenomena at times other than during total solar eclipses. (Eclipses
do not happen very often and one usually has to travel quite far to observe
them.) Almost simultaneously, Janssen and Lockyer developed the prominence telescope for this purpose. They knew that the light of the prominence
is considerably different from that of the solar photosphere. The photospheric light is distributed over the whole spectrum, while the prominences
only radiate at discrete wavelengths. If one considers the spectrum as

* Young's bright green line is due to iron atoms which have been stripped of 13 outer electrons
  (Fe^{13+}). (Tr. note)

Fig. 29. J. N. Lockyer at his telescope/spectroscope

viewed through a high dispersion spectroscope in a region which corresponds to the wavelength of the prominence light, the prominence light is observed practically at full intensity while the other sunlight appears greatly dimmed. Moreover, the lines of the solar spectrum are dark and those of the prominences are bright, such that the prominence spectrum can even be observed inside the solar limb. Under the direction of Huggins and Zöllner this method was so well perfected that not only did the prominence spectrum become visible, but also the shape of the prominence became visible in the spectroscope.

After the 1870s the desire to photograph the whole solar disk in monochromatic light led to a series of investigations which finally resulted in the

Fig. 30. Solar prominences, observed by Lockyer on 14 March 1869

spectroheliographs developed by G. E. Hale (1893) and A. Deslandres (1894). Such an apparatus uses two slits which are adjustable. One of the slits traces over the solar disk and produces a spectrum in combination with a grating. The other slit isolates the very line in whose light the Sun is to be photographed. It traces over the surface of the plate, such that the solar disk is built up in the light of a spectral line. Later the spectroheliograph underwent various improvements. Today extremely narrow-band interference filters are used in order to obtain pictures of the Sun in the light of selected wavelengths.

The study of the solar limb regions soon showed that the Sun is surrounded by a gaseous envelope primarily made of hydrogen. Because of the colored lines in the spectrum of this region, Lockyer christened it the 'chromosphere'. The prominences are expulsions of gaseous material which tower above this chromosphere and extend into the corona. The chromosphere was correctly regarded as a layer of the solar atmosphere which is found right outside the photosphere but is substantially greater in extent.

Scientists of different disciplines were stimulated by the discovery of helium and other findings. They consequently resolved to investigate solar phenomena carefully. For example, in 1872 C. A. Young published a

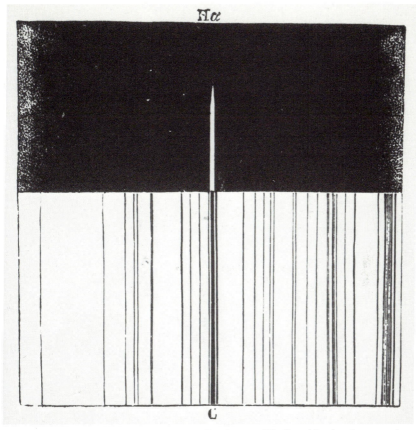

Fig. 31. Emission and absorption lines of hydrogen (Hα lines) in the chromosphere of the Sun and on the solar disk

list of 273 lines in the chromospheric spectrum (flash spectrum). The later application of astrophysics for studies of the different regions of the solar atmosphere led to still more comprehensive results. In 1930 more than 3000 lines were known in the chromospheric spectrum alone.

Sunspots, which are to be counted among the oldest known phenomena, also inspired careful observations. These formations which appear suddenly and then disappear were carefully monitored during the entire pre-spectroscopic period of solar research. Concerning their nature, however, one could only speculate. One of the widespread beliefs was the theory of William Herschel, who regarded the sunspots as openings in the atmosphere through which one saw the dark solar core.

There were apparently good reasons for this. In the eighteenth century A. Wilson had already proposed the observationally based theory that the sunspots are funnel-shaped depressions in the solar surface. Starting on

Fig. 32. Sunspot group on 5 June 1864, drawn by Nasmyth

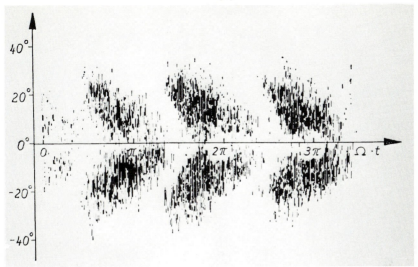

Fig. 33. 'Butterfly diagram' from sunspot observations of 1874–1913. The diagram shows the distribution of sunspots as a function of solar latitude

22 November 1769 he followed a large sunspot with a markedly pronounced umbra and penumbra over a whole rotational period. He thereby determined that this circular formation near the middle of the Sun experienced systematic changes of form with its approach toward the limb – the center grew steadily smaller whereas the 'halo' slowly decreased in width, but on the western side only. Eventually the spot was completely foreshortened at the limb while the 'halo' could still be seen as a tiny streak. With the reappearance of the long-lived formation on the eastern solar limb after 13 days the very same phenomenon was again to be seen; this time, however, the width of the eastern part of the penumbra grew faster than the center. Wilson regarded this appearance as a perspective effect of funnel-shaped spots whose centers lie deeper than the 'halos'.

The real scientific results were limited to laws concerning the rate of appearance of sunspots. In 1843 S. H. Schwabe discovered that the frequency of appearance of sunspots displays approximately a 10-year cycle. R. Wolf proved in 1852, through study of sunspot observations of the previous 242 years, which he put together from widely scattered sources, that this is a general law; he derived a sunspot period of 11.1 years. In 1850 Wolf also introduced the 'relative numbers' still used today for sunspot observation.

Extremely surprising correlations between solar phenomena and facts about the physics of the Earth were immediately discovered from this. In the nineteenth century terrestrial magnetism was the subject of very extensive investigations. Never before were there so many experiments and so much theoretical work concerning the Earth's magnetism as at that time. A. von Humboldt energetically worked on a large-scale magnetic survey of the Earth and with this established the foundation of international geomagnetic observations. Following T. Mayer, L. Euler, and C. Hansteen, it was Gauss, in retrospect, who accomplished the greatest achievement of theory concerning terrestrial magnetism; this was laid down in his classical work *General Theory of Terrestrial Magnetism* (1839). J. von Lamont was busily involved in the area of terrestrial magnetic measurements; he determined the magnetic constant at a number of places in Europe and for many years monitored the daily and yearly variations of terrestrial magnetic strength.

In 1852 Wolf, Sabine, Gautier, and Lamont noted a correlation between the appearance of sunspots and the terrestrial magnetic data. The discovery of the relationship between these two celestial bodies situated so distantly from each other led to the establishment of a whole new area of research which deals with solar–terrestrial relationships.

A new epoch in sunspot research also began with the use of spectroscopy. In 1866 Lockyer carried out investigations of sunspot spectra and for the first time studied particular areas of the solar surface with a spectroscope. Undoubtedly, it was his intention to determine with these investigations something of the nature of sunspots. Actually, the results were promising, for the spectra of sunspots were very similar to the spectrum of the photosphere. Huggins even maintained that they were completely identical. With this it was clearly shown that sunspots must also be gaseous formations, which refuted the theory of Herschel. However, a valid scientific theory could not yet be worked out from the results. Rather, various scientists had greatly differing opinions on this matter. Some, like Kirchhoff and Spörer, considered the sunspots to be formations in the extended solar atmosphere. Similar to the phenomena in the Earth's atmosphere, sunspots might also be local temperature differences and, consequently, cloud formations. S. H. Schwabe and Secchi considered the sunspots to be holes in the photosphere; Zöllner spoke of 'scoria'.

Toward the end of the nineteenth century there was still no valid theory of sunspots; also lacking was a satisfactory reason for their 11-year cycle. Some astronomers sought to explain sunspots as an 'ebb and flow effect' of different planets on the Sun. How far we still were at that time from a solution of the problem and how complicated the processes were in actuality soon became clear from new results – the observation of split lines in the spectra of sunspots. This was first observed in the nineteenth century. G. E. Hale interpreted this phenomenon in 1908 as the Zeeman effect and proved that the sunspots are areas of strong magnetic fields. The detailed investigations which followed this discovery were described in 1938 by Hale and S. B. Nicholson in the comprehensive work *Magnetic Observations of Sunspots*. The important, well-known modern laws of sunspot magnetism, in particular the law of bipolar spot groups as well as the law of the magnetic period being equal to two sunspot cycles, are contained in this work.

On the basis of the form of the solar corona it was assumed early on that the Sun also has a general magnetic field. The measurements which Hale carried out after 1913 also seemed to substantiate this, but the line splittings due to the Zeeman effect were so small that he could not be certain.

In order to be able to study this phenomenon without having to wait for a total solar eclipse, beginning in the 1880s solar physicists endeavored to construct an apparatus which permitted the observation and photography of the solar corona at the same time. However, due to the uncertainty of the data concerning the corona, a successful model could not be built right away. Not until 1930 did the Frenchman B. Lyot succeed in constructing a usable coronagraph. It produces an artificial solar eclipse in the telescope.

In 1909 J. Evershed began measuring the Doppler shifts of weak Fraunhofer lines in the solar spectrum. This also helped clarify our understanding of the nature of sunspots. He derived the flow relationships in the sunspots and was able to show that the deeper layers of sunspots represent inflow of gaseous material and the higher layers represent outflow. The more detailed investigation of these phenomena soon led to the conviction that the sunspots were powerful vortices of photospheric material.

The extraordinarily large amount of data made a definitive theory of sunspots a very difficult problem in the first decades of the twentieth century. Besides the theory of single spots (magnetic fields, cooling with respect to the quiet photosphere), it was necessary to explain validly the numerous statistical findings like spot zones, zonal wandering, periodicities and change of polarity. The endeavors undertaken in this area, particularly the contributions of H. N. Russell, E. A. Milne, and H. Alfvén, did not lead to a definitive and general theory.

With substantial collaboration in the past few years M. Steenbeck and his co-workers have succeeded in explaining the 22-year sunspot cycle on the basis of magnetohydrodynamics.

The investigation of the Sun since the development of astrophysics did not only bring about significant knowledge concerning the Sun as an individual star in the cosmos. The role of the Sun as a prototype simultaneously became recognized for stellar astrophysics. Because of the relatively small distance of the Sun it was possible to study numerous problems of general stellar physics using the Sun and to develop important equipment and methods of research to be applied to other stars. This, among other things, proved to be true for a solution of the question of energy production of the stars and temperature measurements (see pp. 121 ff.).

VARIABLE STARS

A close interrelationship exists between the development of photometry and the investigation of stars with variable luminosity, for the measurement of luminosity changes constitutes photometric data. The early development of photometry is due in part to the interest in variable stars. Only twelve of these objects were known in 1786.

F. W. Argelander is considered the father of variable star research. In 1844 he published an 'Aufforderung an Freunde der Astronomie' ('Invitation to friends of astronomy') in which he called for a systematic observation of variable stars; with it he proposed and pursued the goal of discovering more of such objects. Although nothing was yet known at that time of the significance of variable stars for astronomy, Argelander nevertheless tried to persuade his readers that great enjoyment is derived from those things which bring us knowledge, and he requested that everyone 'do his part for our understanding of the wonderful arrangement of the universe'. One should not be discouraged by the difficulties encountered; finally one should aspire to 'pave the way'[12] for those who come after us. The 'Invitation' also contained a new method for more accurately estimating the luminosities of stars without complicated equipment, the so-called Argelander step method.* The call for action served its purpose; in the ensuing decades many new variable stars were discovered, particularly after the development of photometric measuring devices, along with the beginning of the large surveys and rationalization of work by means of photography (Fig. 34).

As an explanation of the phenomenon of variable stars Zöllner made the suggestion that these were stars in the cooling phase of their evolution. The leftover material arising as a result produces an effect in the rotation, such that the luminosity periodically decreases. This theory was generally accepted for a quarter of a century.

The active search for variable stars began at Harvard College Observatory in 1890, as a result of which the number of known objects rose rapidly. The great differences in the light curves of different variable stars were already known in the eighteenth century. Similar to other data, the classification of the objects was made according to their light curves. This was an important point for the later study of the nature of various types. E. C. Pickering suggested five classes in 1881 which were named according to the prototypes: the Mira Ceti stars, the δ Cephei stars, the Algol stars, as well as the novae and the stars with irregular light variations. This early classification scheme was later expanded according to continuously updated material, whereby the other data such as absolute magnitude or spectral type were used as classification characteristics.

---

* Once a star field containing a variable star has been calibrated by means of photographic or photoelectric photometry, it is a simple matter to estimate visually the brightness of the variable star using the step method. The American Association of Variable Star Observers (founded 1911) is the largest organization devoted to the study of variable stars using this method. Millions of variable star estimates have been made by AAVSO members over the years. Because of the large number of variables known today, only an organization such as the AAVSO, with its worldwide network of observers, can maintain up-to-date light curves for the stars. The results are widely used by professional astronomers. (Tr. note)

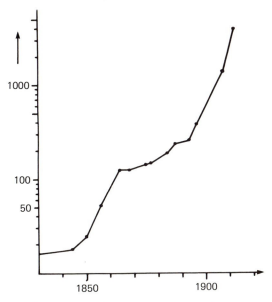

Fig. 34. The number of known variable stars from 1830 to 1910

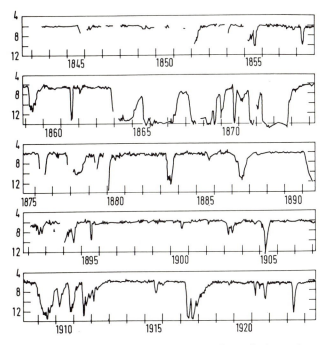

Fig. 35. Light curve of the variable star R Corona Borealis from 1823 to 1924

Spectral analysis of novae brought new insights. In 1866 W. Huggins obtained the first nova spectrum. Such research elicited the greatest attention, for there was no scientific theory for these stars. No one believed anymore that they were stars being born, as Tycho Brahe had assumed. Just as absurd was the hypothesis that novae were stars with extremely non-uniform surface brightness which suddenly turn their luminous sides toward the terrestrial observers. However, there were no new theories.

Huggins then observed that the nova spectrum was comprised of two individual spectra: a marked emission spectrum of luminous gas superimposed on a normal stellar spectrum with absorption lines. Huggins concluded from this that the nova phenomenon had to do with a stellar catastrophe associated with an extraordinary change of state of gas.

When more details were known with improved equipment the theories became even more complicated. The observed Doppler shifts signified an expanding shell of gas.

Indeed, the question still remained completely open as to whether the nova phenomenon had to do with the stellar interior, or, as H. von Seeliger maintained, an interaction with the cosmic dust clouds.

The complicated interrelationships which actively cause a nova outburst could not be resolved entirely until recently. A prospect for their solution in the nineteenth century therefore did not happen, for no information concerning the structure, energy production, and evolution of stars was available.

Greater success was achieved with other types of variables. In 1889 H. C. Vogel succeeded in discovering the double star nature of the variable stars Algol (β Persei) and Spica (α Virginis) on the basis of the periodically doubled spectral lines. The spectrum of each is practically a superimposition of the two spectra of the components, which alternately approach and recede from us due to their revolution about a common center of mass. From the spectra of the Algol system Vogel and Scheiner derived the orbital velocity of the brighter component, and, with the use of the elements of the light curve as well as plausible assumptions, even derived the dimensions of the system and the sum of the masses.

With this the area of research of spectroscopic double stars was opened. At the same time the mystery of the light variations of Algol was cleared up; this was a variable star which had been known since 1669 and whose variability could not be explained for the 220 intervening years. It was an eclipsing variable. The cause of light variation lay in the mutual orbiting of the components, whereby the brighter star sometimes hides the fainter one and the fainter one sometimes hides the brighter one.

In quick succession came the discovery of more spectroscopic double stars. Consequently, these stars could be distinguished from the 'physical variables', for the cause of their light variation was purely a geometrical effect. The physical variables still remained unexplained. They were in general regarded as rare exceptions which could in no way be considered normal stars. This opinion, however, was revised after the introduction of the Hertzsprung–Russell Diagram (see pp. 123 ff.), for now it appeared that the different types of variables occupied completely different regions of this diagram. The importance of the diagram as a snapshot of evolution – already interpreted as such at that time – necessarily led to the notion that the variables represent specific stages of evolution of the stars. That was a crucial realization. Already in 1865 Zöllner had thought that a nova outburst constituted a monumental and conditional stage in the life of a star. However, the greatest difficulties still remained, like the significance of the facts established by astrophysics concerning variables. The great success concerning the interpretation of the light curves of spectroscopic double stars let one assume that significant information could be obtained for all variables. In 1894 Belopol'skiĭ measured periodic variations in the Doppler shift of spectral lines for a variable and found that this period matched the period of light variation. In 1916 H. Shapley and W. S. Adams found that systematic variations in the spectrum also appear in connection with the luminosity variations. At the time of maximum luminosity the very same lines appear very distinctly which correspond to high temperatures, and at minimum the lines which correspond to lower temperatures appear stronger. However, the assumption that these stars were spectroscopic double stars had to be reconciled with two other facts: (1) only one spectrum was visible (the second component was missing); and (2) according to the Doppler interpretation the fastest approach of the surface of the star was observed at luminosity maximum and the fastest recession was observed at minimum. Nevertheless, the line shifts were in general regarded as the superimposition of the spectra of two components revolving about each other, and many authors, among them A. W. Roberts, K. Schwarzschild, and H. D. Curtis, adhered to *ad hoc* hypotheses in order to justify the assumption that they were double stars which happened to be variable.

H. Ludendorff corrected this error when he noted that the observed Doppler shifts could possibly be due to other causes; they did not necessarily imply a double star nature for the objects. The subsequent study of this idea by Shapley then led to the pulsation theory of physical variables, according to which these stars are unstable masses of gas which exhibit oscillations, such that the observed Doppler shifts are due to the radial motions of the masses of gas of the periodically swelling and shrinking star.

In 1918 Eddington mathematically worked out the pulsation theory and with it laid the cornerstone for a modern theory of physical variables.

The study of variable stars is an important area of research of modern astrophysics, for the variability is unequivocally an expression of a phase in the evolution of these objects.

COMETS AND NEBULAE

Astronomers gathered fundamental information on comets with the help of spectroscopic methods. Previously we had only celestial mechanical data. Nothing was known for certain concerning the nature of comets. From 1800 to 1813 in the *Monatliche Correspondenz zur Beförderung der Erd- und Himmels-Kunde* alone more than 400 works appeared dealing with cometary positions and orbits as well as procedures for orbit calculation. During the same period of time, on the other hand, this journal published only six very speculative essays concerning the nature of comets.

The more information that was acquired with the ever more powerful telescopes, the more apparent became the lack of precise models of comets. At the time of the appearance of Halley's Comet in late autumn 1835, even Bessel – so cautious in this regard – expressed thoughts concerning the nature of comets.

In particular, he was interested in the development of the comet's tail and the source of the comet's luminosity. When F. Arago found that the light of comets is partially polarized, it signified that the comet reflects sunlight and also gives off some of its own light. Other details were entirely left open to speculation.

The first comet visible after the introduction of spectral analysis appeared in 1864. Huggins, Secchi, and Donati directed their spectroscopes toward the object with highest expectations. Primarily they found the well-known continuous solar spectrum. Emission lines were also visible, signifying a gaseous component. However, this said nothing concerning the nature of the reflecting material of the comet.

The spectroscopic observations of comets in the next few years led to a series of detailed findings; however, they were not sufficient for deriving a satisfactory model of the physical nature of comets. The spirited laboratory investigations which J. Tyndall began in 1869 also did not provide any unequivocal answer to this question. In particular, there were no sufficiently bright comets observed during those years to aid in the clarification of the comet problem.

The theories which were nevertheless developed regarding comets, by W. Zenker and J. K. F. Zöllner, among others, were the subject of a lively scientific debate.

The Russian astrophysicist F. A. Bredikhin succeeded in detailing the first real scientific theory of comet tails. He assumed – as did Zöllner later – that some kind of solar influence causes the growth of the tail, and that the comet represents an accumulation of dust and gas which at the same time is regarded as the source of material for the corresponding tail when the comet is near the Sun.

Investigations of comets rapidly gathered momentum during the next appearance of Halley's Comet in 1910. On the one hand this comet was an

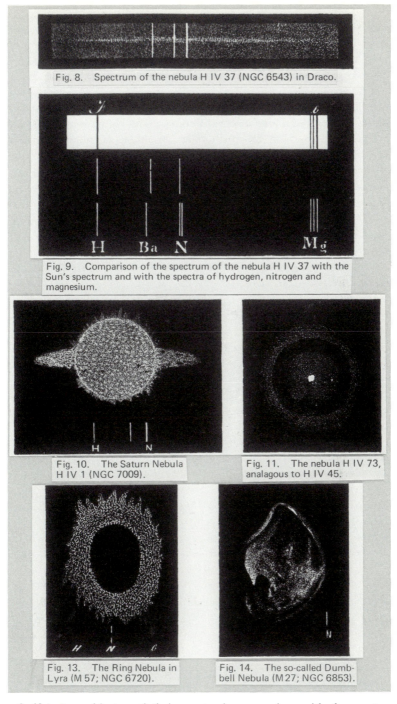

Fig. 8. Spectrum of the nebula H IV 37 (NGC 6543) in Draco.

Fig. 9. Comparison of the spectrum of the nebula H IV 37 with the Sun's spectrum and with the spectra of hydrogen, nitrogen and magnesium.

Fig. 10. The Saturn Nebula H IV 1 (NGC 7009).

Fig. 11. The nebula H IV 73, analagous to H IV 45.

Fig. 13. The Ring Nebula in Lyra (M 57; NGC 6720).

Fig. 14. The so-called Dumb-bell Nebula (M 27; NGC 6853).

Fig. 36. Nebulous objects and their spectra in comparison with the spectra of luminous gas, according to observations of W. Huggins

excellent object for spectroscopic studies; on the other hand the astrophysical methods were greatly improved in the meantime. Here for the first time
all astrophysical methods were applied to a great extent. With the very
comprehensive scientific considerations, which had up till then only concerned themselves with the search for comets, a new epoch of comet
research began which was substantially furthered by the new discoveries of
physics. It led to the fundamental finding that a comet represents a frozen
mass of meteorites which is enveloped in a shell of gas and which undergoes
a continual process of breaking apart.

Formerly regarded superstitiously as harbingers of doom, comets have
since been shown to be members of our solar system, but we have yet to solve
numerous problems concerning them owing to their complex nature.

Concerning the nature of the numerous diffuse, nebulous formations in
the sky which appear in a great variety of forms, almost nothing was known
for a very long time. Following in his father's footsteps, John Herschel
compiled a catalog of 5097 nebulous objects. This led to the *New General
Catalogue* (*NGC*) of Dreyer (1888). At the beginning of the twentieth century
approximately 8000 nebulae were known. With the growing capabilities of
telescopes, more and more of these objects were shown to be composed of
stars. As a result the question of whether this was true for all nebulae came
about. Was it only a question of still larger telescopes, or do 'genuine'
nebulae exist in the cosmos?

Huggins directed his spectroscope toward a nebulous object for the first
time in the summer of 1864. The spectrum consisted of three distinct bright
lines. Huggins could determine from this 'that it is not a cluster of individual
stars, but is an actual nebula'.[13]

This finding alone was very significant. With it the existence of a new type
of material in the universe was unequivocally demonstrated, though previously there were speculations concerning its existence. On one hand this
discovery was easily reconciled with the cosmogonical models of Kant and
Herschel; on the other hand it gave credence to speculations concerning
uniformly distributed, light-absorbing material (see pp. 142 ff.).

Today the study of interstellar matter is a separate area of research within
astronomy. The investigations associated with this material are extremely
wide-ranging: they deal with questions of cosmic distance scales to cosmogony, and more recently even involve biological evolution, as the recent
discoveries of complex organic molecules in interstellar space have shown.

3

Microcosmos – macrocosmos

While astronomers were working out the details of new data acquisition methods and were compiling great lists of astronomical data, revolutionary things were taking place in physics. The disparity between observational facts and apparently solid theories became painfully evident over the course of the years. This led to a new ordering of physics. On the one hand, physicists achieved great successes with the elaboration of important questions concerning the world of the atom; on the other hand, the classical notions of space and time were overthrown. The great revolution in physics at the beginning of the twentieth century had been in the making for decades. It had wide-ranging significance for astronomy, and astronomy helped further clarify many actual problems. 'The road to a knowledge of the stars leads through the atom, and important knowledge of the atom has been reached through the stars',[1] wrote the pioneer of stellar physics A. S. Eddington.

Of particular importance for astronomy was the investigation of electromagnetic fields and the study of their origin, for astronomy is primarily concerned with objects that radiate.

RADIATION AND ATOMIC THEORY

The question of radiation laws has been a prime area of concern of physics since Kirchhoff's time.

In 1860 Kirchhoff showed that the relationship between absorbing and emitting power is the same for all bodies; it depends solely on the temperature of the body and the wavelength of the radiation. The amount of radiation corresponds to the emissive power of a 'black body' at this temperature. Kirchhoff defined a black body as one with an absorptive power of 100 per cent.

A new scientific program was called for because it was necessary to find the function which describes the emissive power of a black body, i.e., how the intensity of black body radiation depends on temperature and wavelength. In 1878 J. Stefan found that the total energy of a black body is proportional to the fourth power of the temperature. In 1884 L. Boltzmann provided a theoretical understanding of this relationship (the Stefan–Boltzmann law).

Nothing was known concerning the question of energy distribution, but progress was made in the development of radiation-measuring devices and the most accurate possible experimental model of black bodies. Because of the well-developed lighting industry in Germany, greatly improved light-measuring equipment was designed and built.

The measurement of the energy distribution of black bodies led in 1893 to what is known as Wien's law, which states that as a body is heated up, the peak of the radiation curve shifts toward shorter wavelengths (blue stars are hotter than red stars).

However, this important law for astronomy still did not allow one to specify how the energy $[B(\lambda, T)]$ behaves as a function of wavelength (λ) and temperature (T).

In 1896 W. Wien found a mathematical equation which agreed well with numerous measurements over a short wavelength range. However, the measurements of H. Rubens and F. Kurlbaum showed that the formula did not fit at higher temperatures and longer wavelengths.

After 1896 Max Planck also concerned himself with the question of black body radiation. On the basis of his own calculations he confirmed Wien's law. Planck then learned of the measurements of Kurlbaum and Rubens and attempted to combine those results with the earlier ones and to describe them by a mathematical expression. On 19 October 1900 Planck presented his radiation law at a session of the Physical Society in Berlin. Rubens tested the law the following evening according to his own laboratory data and communicated the complete agreement between the formula and experiment to Planck the next morning. With this the question of the energy distribution in the spectrum of a black body was explained by the Planckian radiation law. However, Planck was not yet satisfied with this formula, for 'if one assumes its absolute, precise validity, the radiation formula would only represent a lucky guess rather than simply a formal explanation'.[2]

Planck himself set out to 'find an actual physical explanation'[3] of his formula. The result was a revolution in physics; the radiation formula led to the development of a quantum mechanics – a new representation vastly different from classical physics. Energy is not radiated in any and all amounts, but always in little 'packets' only, the so-called quanta. The energy of these quanta is determined by the frequency of the radiation and by a universal natural constant referred to as 'Planck's constant'.

On 14 December 1900 Planck presented this result before the Physical Society in Berlin. The paper which was published shortly thereafter was entitled 'Zur Theorie des Gesetzes der Energieverteilung im Normalspektrum' ('On the theory of the law of the energy distribution in a normal spectrum'). The basis of modern quantum theory was thereby established.

Planck had completely disregarded the question of exactly how the radiation originates. He never explained the role of atoms in the radiation process. Moreover, his arguments are based on the notion of the 'simple harmonic oscillator'. After Einstein's light-quantum theory and also his theory of specific heat (1907) had demonstrated the great significance of the Planckian quantum hypothesis, its consequences were amplified to a great extent when Niels Bohr combined the quantum hypothesis with the models of the constitution of atoms.

The dispute over the existence of atoms (the smallest chemical constituents of material, unable to be broken down any further) had been going on for centuries. At the beginning of the nineteenth century Dalton's law of multiple proportions substantiated belief in the possible existence of atoms. The kinetic theory of heat, as it had been developed in the second half of the nineteenth century by Clausius, Krönig, Maxwell, and Boltzmann, could explain such fundamental phenomena as the increase of gas pressure with temperature, keeping the volume constant. The Faraday equivalence law of electrolysis is also in accord with the model of a discrete structure of material.

However, the historically most important clues concerning the atomic structure of matter came from the investigations with discharges in vacuum tubes. The experimental physicist P. Lenard had shown in a series of important experiments that atoms are mostly empty space. He was also able to detect the small particles of the cathode rays outside the vacuum tubes after they had penetrated a 3-μm thick layer of aluminium. The atomic dimensions derived on the basis of the kinetic theory of gas showed that in such a sheet of material there were approximately 10 000 atomic layers overlapping each other. It was regarded as extremely unlikely that the electrons detected could have passed straight through large gaps.

In 1912 E. Rutherford repeated the scattering experiment of Lenard with more massive alpha particles (helium nuclei). As a result he confirmed and improved on the previous results. Together with his students Geiger and Marsden he accurately determined the deflection angle of the alpha particles for different foil materials and velocities and found that a few particles were deflected from the foil with angles of up to 150 degrees from the direction of incidence. Rutherford concluded that these extremely large deflections are due to strong electric fields which are confined to a small volume. From the comparison of corresponding measurements of the scattering from different foils he found that the deflection increases with the atomic weight of the scattering element. It became clear that the charges (expressed in units of the elementary electron charge e) precisely agreed with the atomic number of the scattering element in the period table. Significant conclusions concerning the structure of the atom could be made from this. The atom must consist of a positively charged nucleus which causes the occasional but large deflections of the likewise positively charged alpha particles. As the atom must appear neutral from afar, a number of electrons corresponding to the nuclear charge (atomic number) must circle the nucleus. This interpretation was first expressed clearly in 1913 by van den Broeck and was substantiated by findings from chemistry. According to the interpretation the electrons do not fall into the nucleus; they move about it like the planets about the Sun, i.e., in elliptical orbits according to the Keplerian laws. This was the 'planetary model' of the atom as Rutherford and his colleagues presented it.

However, one cannot overlook the fact that this model leads to a number of theoretical and experimental facts which are incompatible. According to the laws of classical electrodynamics the electron revolving about the nucleus constitutes a dipole and would continuously radiate energy in the form of electromagnetic waves. As a result a stable atom would not be possible; the electrons would spiral into the nucleus after a few revolutions. Besides, spectroscopy pointed to the existence of sharp lines whose origin remained completely uncertain.

The Danish physicist Niels Bohr, then 27 years old, concerned himself with these questions urgently requiring solution. He had himself participated in the experimental foundation of the 'planetary model' with Rutherford in Manchester, where he had visited as a guest. Within a few months, in early 1913, Bohr obtained a solution. Under the influence of J. Stark's book *Principles of Atomic Dynamics* (1911) and with consideration of the well-known facts concerning spectral lines, Bohr hammered out his own atomic model which explained the stability of atoms as well as the origin of spectral lines. His principal accomplishment was that he combined the already existing models of the atom with the concept of energy quanta. Bohr postulated that there are specific electron orbitals in the atom, measured by their energy, which the electrons occupy according to classical theory, and in which they move about without radiating. The emission of light comes about when the electrons 'spring' from one orbital to another of lower energy, where the energy difference corresponds to the particular frequency of the electromagnetic wave. In his classical work *On the Constitution of Atoms and Molecules* (1913) Bohr laid the foundation of quantum mechanical atomic theory. Further improved by A. Sommerfeld, quantum theory developed from it. The classical work of that epoch of atomic physics is Sommerfeld's book *Atomic Structure and Spectral Lines* (1919). The cause of spectral lines, which had been sought since the time of Kirchhoff and Bunsen, was thereby found.

THE INTERPRETATION OF STELLAR SPECTRA

These achievements of theoretical and experimental physics were of the greatest value for the understanding of spectra of celestial bodies, especially stars.

At first the great differences in stellar spectra were understandably thought to be due to differences of chemical composition. Not until the development of atomic theory and its application to stellar spectroscopy did a fundamentally different point of view arise.

In 1920 the Indian astrophysicist M. Saha achieved a most significant result for the physical explanation of stellar spectra. Saha started with the Bohr–Sommerfeld atomic theory and considered the ionization processes affecting the atoms at the surface of a star, given the effective temperature. He succeeded in deriving an exact equation relating temperature, pressure, and degree of ionization for the different atoms. The original work of Saha – often quoted today – seemed too speculative to the editor of the *Astrophysical Journal* and it was flatly rejected for publication. It then appeared in 1920 in the *Philosophical Magazine*. However, the Saha ionization theory has been splendidly confirmed and has contributed substantially to our understanding of the physical conditions in stellar atmospheres as derived from complicated stellar spectra. On the whole Saha's theory created the foundations of a scientific study of stellar atmospheres. In 1925 Cecilia H. Payne synthesized Saha's theoretical derivation and the observational data in her book *Stellar Atmospheres*.

One of the earliest and most convincing successes of the theory was the fundamental explanation of the relative abundance of elements in stellar atmospheres. The surprising result was that the majority of stars proved to have essentially similar compositions, and that the differences in the spectra were solely due to different pressures and temperatures. The hot atmosphere of a B star, for example, contains a considerably greater fraction of ionized atoms in comparison to the cool atmosphere of a K star. So, on the basis of the Saha theory, a whole series of previously unidentified lines in the spectra of stars could be identified; these belonged to known elements ionized to a great degree. The strong lines of ionized metals which exist in the spectra of B stars belong to the same chemical elements as the lines appearing in other spectral regions in the K stars, for they show up there as normal atomic spectral lines of neutral metals.

According to the Harvard spectral classification, for example, the A to G stars are differentiated by the decreasing intensity of the hydrogen absorption lines. The Saha ionization theory shows that this is not to be interpreted as a decreasing hydrogen abundance, rather that the degree of ionization of hydrogen increases with increasing temperature from K to A, and therefore a greater number of hydrogen atoms experience the condition necessary for emission of the Balmer lines of hydrogen.

Thus the Saha ionization theory succeeded in giving a physical foundation to the spectral classification while it proved that the internationally adopted spectral sequence at the same time represents a temperature sequence with decreasing temperature from the 'early' to 'late' spectral classes.

More than anyone else, R. H. Fowler and E. A. Milne contributed to the subsequent development of the theory of stellar spectra. The result of these applications of atomic physics to stellar spectra was a great stimulus to stellar spectroscopy, in particular the theory of stellar atmospheres. At the same time this led to a profound understanding concerning the nature of certain phenomena in the universe.

The investigation of the spectra of galactic nebulae is another application of atomic theory to the interpretation of spectra of celestial bodies. The theory of emission in gaseous nebulae is based on the assumption that the emission results from ultraviolet light from a nearby star. However, in addition to the expected lines (e.g., the emission lines of the Balmer series of hydrogen), several lines were also found in the spectra of nebulae which could not be shown to have been caused by known elements. Because these green lines are found in nearly all nebular spectra, the idea was proposed, analogous to the discovery of helium, that they were caused by a previously unknown element which is to be found only in these nebulae. It was given the name 'nebulium'.

However, certain findings in stellar spectroscopy led to scepticism with regard to the nebulium hypothesis: (1) there were strong indications that the same elements are found everywhere in the universe; and (2) in 1925 only a small number of observed lines in stellar spectra required explanation.

Then in 1927 I. S. Bowen and R. H. Fowler solved the riddle of the nebulium lines, not by obtaining laboratory spectra of some substance, but by means of theoretical considerations. From the Bohr atomic theory it was known that the duration of excitation for normally excited atomic states is very short; after approximately 10^{-8} seconds an electron in an excited state 'springs' back to the ground state, emitting one or more photons (depending on how many energy levels it pauses at on the way down). Besides this there are also levels of excitation in which the electron lingers a while longer, the 'metastable states'. Here a transition to states of lower energy is in general not allowable; it is forbidden by certain 'selection rules'. The metastable states therefore do not in general lead to light emission, but rather the energy is transformed by means of collisions with other atoms. However, such collisions happen so seldom in the extremely rarified galactic nebulae that the metastable states lead to emission of 'forbidden lines'. In particular, the 'nebulium lines' turned out to be 'forbidden lines' of the known elements nitrogen and oxygen. The lines of 'coronium' in the solar corona were similarly explained.*

---

* 'Coronium' lines were found to be due to allowed (*not* forbidden) transitions of electrons of highly ionized atoms of iron, nickel, and chromium. There are still a few coronal emission features without positive identification. (Tr. note)

THE TEMPERATURES OF THE STARS

Until the year 1900 the whole theoretical foundation of astrophysics was manifested in Kirchhoff's law of black body radiation, in the Doppler principle, and in the gas laws. These relations did not suffice for a determination of stellar temperatures. The derivation of the surface temperature of the Sun was only achieved on the basis of the solar constant and with the application of the Stefan–Boltzmann law (see pp. 98 ff.).

Wien's shift law was applicable only to a limited degree, for the intensity maxima corresponding to temperatures over 7000 degrees are at such short wavelengths that the measurement of the maximum could not be made with equipment available at that time. Also, the Stefan–Boltzmann law could not be used because we did not have a sufficient understanding of the radiation processes in stellar atmospheres. Therefore, it was the Planckian radiation law that first made possible the temperature measurement for all stars. Besides numerous experimental problems which had to be solved, the role of atmospheric extinction (i.e., the dimming of starlight as a function of wavelength in the atmosphere) also had to be accurately investigated. J. Wilsing, among others, studied this problem.

In 1873 Zöllner had suggested that the information concerning the intensity distribution in the spectrum of the stars would permit 'at least the relative temperatures to be qualitatively determined, i.e., to decide which of two stars has a higher temperature'.[4]. In 1880, on the basis of a series of fundamental spectrophotometric measurements carried out under difficult conditions, Vogel made the statement that the red stars approximately correspond to the temperature of electric arcs (approximately 4000 K). He was also convinced that it would be possible exactly to derive stellar temperatures by means of the Kirchhoff function. Vogel thereby paved the way for success. Other investigations, however, did not succeed; for example, the attempt to derive temperatures on the basis of the appearance of certain laboratory lines. Still, the toilsome spectrophotometric investigations, which Vogel and Müller carried out in 1880 and which had led to qualitative statements concerning stellar temperatures, served in 1901 as a foundation of an exact derivation of stellar temperatures. As a result B. Harkanyi determined the temperatures of the Sun, Sirius, Vega, Aldebaran, and Betelgeuse.

In 1900 the use of the Planckian radiation law for the determination of temperatures of the stars involved the fact that even more considerable experimental difficulties were to be encountered in the carrying out of spectrophotometry. Again it was the Potsdam astrophysicists J. Wilsing and J. Scheiner who in 1909 obtained a pioneering result in this area. They compared selected portions of stellar spectra with the corresponding regions of the comparison spectrum. This alone entailed certain tricks. However, after that the physical problems had to be solved in order to bring the intensity distribution in the spectrum of the comparison source into accord with that of a black body at a particular temperature. The result of the extremely toilsome work was a determination of the temperatures of 109 bright stars. The color temperatures measured in this way were a substantial contribution to the development of methods of temperature determination.

Later Rosenberg worked out a photographic method for temperature determination, in which the intensity distribution in the spectra is measured spectrophotometrically. However, as a result it was shown that the same spectral types gave greatly different temperatures according to the method used. Above all A. Brill concerned himself with the derivation of a uniform temperature scale. He thereby made an important suggestion – to determine the color temperature of the stars from the color index (CI) introduced by K. Schwarzschild. The index gives the difference of luminosity of the object to be investigated in two different spectral regions and is a measure of the intensity distribution. This is relatively independent of discontinuities of the intensity distribution. However, such discontinuities unavoidably arise in stellar spectra, for stars are not ideal black bodies.

The practical realization of the measurements took the form of relative determinations, i.e., the spectra of individual stars were compared with each other and the standardization to absolute values was then achieved in the laboratory. This standardization is the most complicated part of stellar temperature determination, above all because of the great difference between the temperature of the stars and that of a black body.

The problem of stellar temperature determination had not been completely resolved even with this. Moreover, different methods are possible for a determination of temperature which usually lead to somewhat different values. Besides the color temperature (the first to be determined), with the application of other methods one speaks of the so-called gradation temperature, the radiation temperature, and others. Of greatest interest for

Table 6. *Accepted temperature scale (1940)*

Spectral type	Temperature (K)
Bo	22 000
Ao	13 500
Fo	8 550
Go	5 800
Ko	4 370
Mo	3 240

astrophysics is the synthesis of these different temperatures into a standardized temperature, the so-called effective temperature. A star manifests the effective temperature T_e by definition when any unit surface emits the same total energy as a black body of temperature T_e. Often it is necessary to know the surface luminosities of stars and to derive the effective temperatures from this information. Only a small number of stars have been used for this (the Sun, a few eclipsing binaries, and a few stars with interferometrically determined diameters). The scale of effective temperatures is therefore still very uncertain today. In 1940 W. Becker empirically found a relationship between color temperature and effective temperature using Cepheids and eclipsing binaries. The conceptualization of this relationship improved the astrophysical interpretation of color temperature such that it can now be directly used for effective temperatures which serve as the foundations for radiative measurements.

THE HERTZSPRUNG–RUSSELL DIAGRAM AND THE EVOLUTION OF THE STARS

With the growing data base of stellar parallaxes it became possible to calculate another important descriptive element for the stars – their absolute magnitudes. That the stars shine with different brightnesses is connected with their different distances. As a result the apparent luminosities are not a valid descriptive element. However, if one reduces them to a standard distance, so that the 'absolute magnitudes' thus obtained represent proper descriptive elements of the stars, they are a measure of the luminosity and therefore of the energy radiated by the stars per unit time.

Mädler had already concluded from observing the differences in the magnitudes of the components of double stars that the absolute magnitudes of stars also had quite a range. To a great extent this question could only be investigated when sufficiently accurate stellar distances were known through extensive parallax programs. Thus it was that we did not learn until 1900 of the great scattering in the absolute magnitudes of the stars.

At the same time questions of spectral classification were also at the focal point of interest of the astrophysicists. It had been thought from the beginning that the spectra could tell us something of how the stars evolve. It had been assumed that the stars undergo a cooling process over the course of their lifetimes and as a result their colors change from blue to red according to the sequence of spectral classes. The conception of this stellar evolutionary path led to the differentiation of the stars into 'early' and 'late' spectral classes (terms still used today). Now if one assumes that the stars in general have similar structure, then one can expect that the absolute magnitudes of the stars decrease from the 'early' to the 'late' spectral

Fig. 37. Ejnar Hertzsprung

classes. As a result it naturally became of interest to find out if the absolute magnitudes and the spectral classes were indeed correlated.

As early as 1893 the Irish amateur astronomer W. H. S. Monck tried to determine how stellar absolute magnitudes depend on the spectral types. Indeed, at that time the data concerning spectral classes and stellar distances were not extensive or accurate enough for one to reach unequivocal conclusions. When the young Danish engineer and specialist in photochemistry Ejnar Hertzsprung applied himself to the same question more than a decade later the prospects for success were substantially more favorable. In the meantime the spectral classification scheme had been markedly revised by the collaborators at Harvard College Observatory. Something was still left to be desired concerning stellar parallaxes. Indeed, the measurements already under consideration and the important use of proper motions only slightly improved the situation.

The basic idea of Hertzsprung consisted of deriving the stellar absolute magnitudes from the already existing observations of parallaxes and proper motions and of ascertaining the relationship between absolute magnitude and spectral class likewise from the available material of spectral catalogs. In order not to be thrown off track by considering inaccurate data, he limited himself to a relatively small number of authentic, accurate absolute magnitudes.

Hertzsprung published his findings in 1905 and 1907 in a two part publication under the title 'Zur Strahlung der Sterne' ('On the radiation of stars'). Unfortunately, he published this classical work in the *Zeitschrift für wissenschaftliche Photographie* (Journal for Scientific Photography), such that his results only gradually became known to astronomers.

Later Hertzsprung once again published the essential results in expanded form in the *Astronomische Nachrichten*, and this was very important for the dissemination of his ideas. In the intervening time it was shown that the information and procedures described in this relatively obscure work were to a great extent applicable to the study of the nature of the stars; to a great degree the method succeeded, such that it is still used today. In these articles Hertzsprung came to the important conclusion that if the stars are ordered by color (blue, white, yellow, orange, red), this is the same order of their absolute magnitudes (from brightest to faintest).

As a result of this Hertzsprung surprisingly discovered that some stars of spectral classes G to M had greatly different absolute magnitudes in spite of having identical spectral types. This was the discovery that the stars of late spectral types, in contrast to the others, fall into two groups, those with large luminosities and those with faint luminosities. For the bright ones there appeared to be no strong correlation between spectral type and absolute magnitude. Among others, Hertzsprung listed the examples shown in Table 7.

Table 7. *Absolute magnitudes for different stars of the spectral types G to M, according to Hertzsprung*

Object	Spectrum	Magnitude reduced to a unit distance of 1 pc[5]
α Aurigae	G	$-5^\mathrm{m}\!.09$
α Centauri		-0.24
α Bootis	K	-7.10
70 Ophiuchi		$+0.53$
α Orionis	M	-6.47
Lac. 9352		$+4.76$

Although Hertzsprung did not know why there should be two luminosity classes, he already referred to the stars of later spectral types with higher luminosity as *Riesen* (giant) stars.* Indeed, at the end of the nineteenth century W. H. S. Monck already used the term 'giants' in the English form. However, no concrete model was proposed with this nomenclature; it simply explained the large luminosity.

Still, the classical work of Hertzsprung contained a more wide-reaching idea which represents interesting evidence of the role of imagination in scientific research. Hertzsprung emphatically advocated the notion that the great differences in the absolute magnitudes of the stars of late spectral classes must also be discernible in some way or other in the spectra of these stars, although for the classification the standard characteristics agreed. In 1905 there were very few clues giving credence to this supposition, and yet Hertzsprung proposed the basic idea for the determination of spectroscopic parallaxes. This was realized barely ten years later.

The criteria for the determination of absolute magnitudes from stellar spectra were discovered in 1914 by Adams and Kohlschütter. The intensities of certain absorption lines were found to correlate strongly with absolute magnitude. Adams and Kohlschütter calibrated the method with stars of known parallax. As opposed to trigonometric parallax measurements, the method of spectroscopic parallaxes could be used on much more distant stars. By 1935 approximately 6000 stars of all spectral classes were measured according to this new method.

As to Hertzsprung's data, it is interesting to note that the deductions he made were essentially the results of desk work only. Hertzsprung hardly used any material which was not already in published form. Thus he determined what information was concealed in the data of other scientists.

---

* Consider two stars of the same spectral type, but with relative luminosities differing by a factor of 100 (a difference of 5 magnitudes). Because they are of the same spectral type, their effective temperatures are essentially the same. Therefore, each square centimeter gives off the same number of photons. If one star has 100 times the luminosity of the other, it must have 100 times the surface area; its radius would be 10 times bigger. (Tr. note)

What consequences did Hertzsprung's findings have for the understanding of stellar evolution?

In 1909 K. Schwarzschild noted: 'It is highly noteworthy and not predicted by any theory of stellar evolution that these giants are so diverse with respect to normal stars',[6] i.e., that the existence of the newly discovered giants could not be reconciled with existing interpretations.

Hertzsprung had attempted to provide an explanation in which he assumed that a certain percentage of stars falls into each of two parallel series. In both the absolute magnitudes decrease with increasing redness in accord with the generally accepted evolutionary theory. The previously known frequency distribution of the stars should be manifested through time; a star finds itself in each spectral class over the course of its lifetime. Later, however, Hertzsprung did not resurrect these ideas, particularly because subsequent research proved that other models were more valid.

Incidentally, in the already-mentioned original publication of Hertzsprung one cannot find the famous Hertzsprung–Russell Diagram (H–R Diagram). Hertzsprung thought it sufficient to publish the findings discovered by him in the form of tables. Not until 1911 did he graphically represent the relationship between color and luminosity for stars in the Pleiades and Hyades as the first color–magnitude diagram.

The first graphical representation of the relationship between spectral type and absolute magnitude of the stars was given by H. N. Russell in 1913. Russell concerned himself with the very same problem, independently of Hertzsprung's research. He thereby achieved the same results as Hertzsprung. Russell proposed the term 'dwarfs' for the stars of later spectral types with fainter absolute magnitudes as the counterpart of the giants. He was also the person who posed the significant question of what the giants and dwarfs really were, for up till then we had only considered their relative sizes: were the giants of the same consistency as the dwarfs but more massive, or were the masses of the giants and dwarfs approximately the same, while the giants had a much greater extent and therefore an extremely small density?

Russell was able, from careful studies of eclipsing binaries, to show unquestionably that the superluminous red stars have much larger diameters with respect to the red stars of smaller luminosity, while their masses are quite comparable with those of other stars. This finding was for Russell a completely essential clue for the modification of the theory of stellar evolution accepted at that time. He considered the giant phase to represent the beginning of stellar evolution. According to the theory the stars begin their life as objects of very large extent and low temperature, as giant stars of spectral type M. Under the influence of gravity the decrease of the diameter and an increase of the temperature should take place until the star reaches its place in the upper left-hand corner of the H–R Diagram as an object of spectral type B. After the end of the contraction phase the cooling

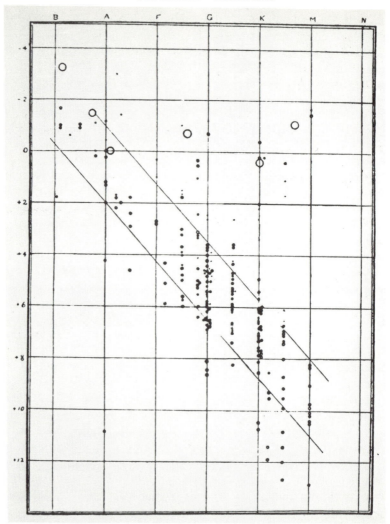

Fig. 38. Russell's first diagram. On the abscissa are given the spectral types and on
the ordinate are the absolute magnitudes

process begins; this again corresponds to a great degree to the previously
advocated interpretation. As a result the star will run through the various
spectral classes, this time in reverse order. It ends as a star of spectral type
M. While it was a gigantic object of small density and bright absolute
magnitude in its youth, in old age it becomes a relatively small object with
faint absolute magnitude.

Russell modified this theory of stellar evolution when he determined that
the masses of the red giants are greater than the masses of other stars. He
based this conclusion on a relatively scanty amount of observational data of

double stars. Similarly, in 1911 Ludendorff found from investigations of spectroscopic double stars that the stars of spectral type B also have substantially greater masses than other stars of the main sequence of the H–R Diagram. Russell then attempted to encompass these two findings into a stellar evolutionary theory. So he came to the conclusion that the massive red giants reach the culmination point of their evolution as stars of spectral type B, while the result of the evolution of less massive young stars is at later spectral classes, i.e., the less massive young stars do not attain the B class at all; instead, the contraction process blends right into the cooling process.

Thus the bulk of data existing at that time led to the assumption that stellar evolution depends on the initial mass.

This theory of evolution has generally been accepted. Subsequent information concerning the characteristics of the stars substantiated it even further. In 1919, from data on double star systems with well-known parallaxes and mass estimates, Hertzsprung empirically derived an equation relating the mass and luminosity of stars; this had already been suggested by J. K. E. Halm in 1911. This equation implied that more massive stars also manifest higher luminosities. This relationship was of very great importance for other areas of modern astrophysics; if it were strictly true, then one could estimate stellar masses directly from stellar luminosities.

This relationship meant that the masses of stars along the main sequence monotonically decrease from the upper left to the lower right. This could only be evidence in favor of the theory of evolution accepted at that time – that one is able to account for a permanent energy production only by means of a permanently decreasing stellar mass.

In 1933 B. Strömgren discovered a new interpretation of the H–R Diagram corresponding to the theoretical state of knowledge at that time. He considered the mass–luminosity relationship (in the meantime theoretically substantiated) and the uniqueness theorem of stellar structure of Vogt and Russell. Strömgren also proposed the term 'Hertzsprung–Russell Diagram' generally used today in order to honor the achievements of Hertzsprung and Russell.

Strömgren suggested that the rate of stellar evolution might depend on the mass of the stars and that the detailed study of stars in clusters – for which one assumes a common age – would be suitable to better clarify this question.

In particular, the Vogt–Russell theorem, according to which the evolution of the stars is only possible if the mass of the star or its chemical composition or both change, clearly showed the limits which the established evolutionary theory of stars had attained in 1930, because, for the changes of mass or chemical composition of the stars, the processes naturally have to be outlined by which the energy of the stars is produced. And concerning this there were only speculations. Strömgren therefore correctly stressed in his important treatise, 'On the interpretation of the Hertzsprung–Russell Diagram', that one could not expect final answers to the question of stellar evolution until one had discovered the energy sources of stars (see pp. 196 ff.).

STELLAR INTERIORS

The study of stellar interiors, without which the nature of stars could not be completely understood, encountered one major difficulty: all radiation of stars which we receive comes from an outer layer of each star which is very thin compared to the radius of the star. A direct indication of what is happening deep inside the stars is not available to us – seemingly a hopeless situation. Nevertheless, that the question of the interior structure of the opaque stars could be successfully studied is very closely associated with the achievements of other scientific disciplines. The high temperatures of stars, which were already known before the Planckian radiation law spelled out the exact procedure for temperature measurement, let one assume that the stars could be modeled as spheres in a gaseous state of equilibrium. Then, by mathematical methods and with the application of gas laws something about the hidden interiors of stars could be calculated. The first treatise which dealt with such considerations was published in 1870 by the English physicist J. H. Lane and carried the title: 'On the theoretical temperature of the Sun; under the hypothesis of a gaseous mass maintaining its volume by its internal heat, and depending on the laws of gases as known to terrestrial experiment'. Lane consequently assumed that the known gas laws were also valid for the Sun, and he investigated the solar interior with the tools of the theoretician. He thereby postulated that the weight of the gas (which, if unopposed, would cause the star to shrink) is balanced by the outward gas pressure of the stellar material.

After Lane many other scholars concerned themselves with this question. In 1907 the astrophysicist R. Emden succeeded in a classical way in modeling the inner structure of the Sun in his book *Gaskugeln* (Gaseous Spheres). It is interesting that this wide-ranging book originated in an attempt to enliven the abstract treatment of thermodynamics in high schools through examples of practical application. The treatment of stars as gaseous spheres is another example of the close interrelationship of different disciplines applied to the clarification of astronomical research problems. Here a knot was tied 'whose threads were made up of nearly all areas of physics'.[7]

Above all A. S. Eddington built on the work of Emden. Eddington created the foundations of the modern theory of the inner structure of the stars still accepted today. He succeeded in calculating how the temperature of stellar interiors changes as a function of the mass and radius of the star; this led to theoretical ideas concerning the luminosity of the stars. Next it

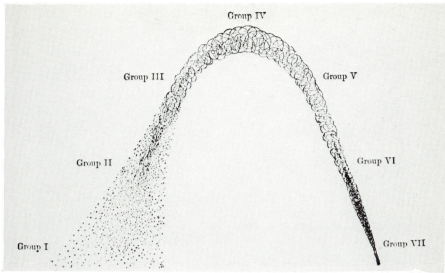

Fig. 39. Lockyer's temperature curve, used as a representation of the evolution of the stars. The different groups represent different temperatures and therefore – in accord with the findings of Lockyer – represent different states of evolution

seemed plausible that the details of the chemical composition of the star are very important for the calculation of the internal temperature, for the temperature can only be estimated if the pressure is known, and this naturally depends on the average density of the particles. However, at the temperatures inside the stars all atoms become ionized. Each atom is thus broken apart; all electrons are separated from the nuclei. Then it was possible to show that the average density of particles is almost independent of the chemical composition of the gas, for heavier atoms also contain more electrons than light atoms according to the nuclear charge. Therefore, a value of 2 was used as the average molecular weight.* A considerable deviation from this value is manifested solely by hydrogen. The hydrogen atom consists of a proton in the nucleus and an electron in orbit. After ionization the average atomic weight is then 0.5. Consequently, Eddington did not worry about the exact chemical composition of the stars, but rather he needed to know the hydrogen abundance. If one assumes a low hydrogen abundance for the stars, a substantially high pressure results and also a substantially high temperature in the stellar interior. Now Eddington assumed a very small percentage of hydrogen. The luminosities derived from the known radii and masses of stars – so the measurements showed –

* The proton is 1836 times more massive than the electron, and the proton and neutron have comparable mass. Thus the electrons are counted as particles in the plasma of the stellar interior, but they contribute very little mass for the calculation of the average molecular weight. (Tr. note)

Fig. 40. A. S. Eddington

always came out too high. This could only have one of two causes: either the actual hydrogen content was much higher than that assumed or the assumed mechanism of transport of electromagnetic waves from the center of the star to the surface was wrong. This factor was also subject to uncertainty. At that time wave mechanics was still in a state of tumultuous development, and no one could say whether the radiation transport within the star up to its surface had been calculated correctly.

At the beginning of the 1930s this question was theoretically and practically resolved. The method of energy transport had essentially been correctly described, such that the differences could only be due to the

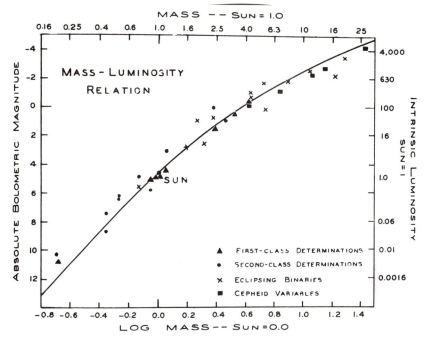

Fig. 41. Mass–luminosity relation of A. S. Eddington
▲ first class determinations
● second class determinations
× eclipsing binaries
■ Cepheids

assumption of an incorrect hydrogen abundance. At the end of the 1920s the interpretation of stellar spectra made possible by the Saha ionization theory demonstrated the enormously high hydrogen content in stellar atmospheres. It was clear that the stars contain a high percentage of hydrogen and, by comparison, a correspondingly small amount of heavier elements.

It was still very difficult to determine how the energy was released in the star as a whole and whether it depended on the temperature and the density, and therefore upon the position inside the star. Eddington had assumed that the energy radiated by the star is the result of a mechanism which could not be more precisely described; furthermore, that the release of energy takes place uniformly throughout the whole star. Other stellar models assumed that the energy is released in the stellar core (the point source model). Because this question could not be decided, this had little influence on the practical use of the stellar models. In each case it was possible to calculate the central temperature from the radius and mass and to determine how the temperature varies as a function of the radius, given the central temperature. Then, by means of the radiative power, one could calculate how much energy streams forth through a unit area at the surface. Multiplying this by the total surface area then gives the luminosity. The luminosity differences which resulted from the different models were very small. For stars with known mass, known radius, and measured luminosity one could even calculate the hydrogen content from the known data owing to the dependence of luminosity on hydrogen content. Finally, Eddington derived the mass–luminosity relation from his theory of the inner structure of the stars. This was very important for astrophysics.

It is amazing how much detailed information one can derive concerning the inner structure of the stars and concerning the interrelationship of the 'internal parameters' with the measurable parameters, without having knowledge of the processes which govern the release of energy.

COSMIC RAYS

The discovery and investigation of cosmic rays is another characteristic example of the interrelationship between the studies of the microcosmos and the macrocosmos.

C. T. R. Wilson, as well as Elster and Geitel, investigated the electrical conductivity of gases in connection with actual physical research. The determination that an electroscope will discharge with time was taken as proof that the air had a non-zero conductivity. Wilson tried to explain this conductivity and in 1900 came to the conclusion that radiation streaming in from the cosmos (waves and/or particles) is responsible for this effect. However, he later abandoned his hypothesis when he found that the effect was essentially the same even when he used lead shielding and carried out the conductivity measurements in a railway tunnel.

Even the radioactivity measurements with ionization chambers showed a similar, noteworthy effect. Evidence always indicated some kind of outside radiation which could be removed either by the use of sufficiently thick wall material or by means of highly purified gases.

In 1901 the point of view again became prevalent that this effect could be attributed to radiation from outer space which was able to penetrate material. On the other hand it could also be due to an effect of terrestrial radioactivity, the masonry of the laboratory, or some other cause. In order to eliminate these influences, investigations with ionization chambers were carried out on top of high towers. In 1909 measurements on the Eiffel Tower in Paris showed a minute but significant diminution of the effect of the unknown radiation. This was to be expected if they were really 'Earth rays'. Later this was interpreted to be evidence *in favor* of cosmic rays.

But the results were not generally acknowledged; instead many physicists adamantly argued that this line of reasoning belonged to the realm of fantasy. In Graz a doctoral dissertation was even defended which attempted to prove that rays could not possibly penetrate from space.

V. Hess and W. Kolhörster were among the convinced proponents of the hypothesis of the cosmic origin of the rays. Hess also was the first to send up a balloon in the course of his research. He found a noticeable increase of radiation intensity with an ascent to 5000 meters. Kolhörster confirmed this finding with a balloon ascent to an altitude of 9000 meters. Although these experimental results were accomplished during the years 1910–1913, even ten years later there was still no unequivocal decision. In 1923, likewise with a balloon ascent, R. A. Millikan attempted to refute the results of his predecessors.

Not until 1926 did the idea of the existence of radiation streaming from space become generally discussed. The question of the origin of the radiation led to a series of investigations which were applicable to the relationship between cosmic events and the ionization stream.

The first result of this research was the determination that the intensity of radiation depends on the apparent solar time, such that a solar origin for the radiation was inferred, but contrary to this were investigations carried out during the total solar eclipses of 21 August 1914 and 29 June 1927 which showed no noticeable decrease of intensity of radiation.

Still more disputable was the correlation of the intensity of radiation with sidereal time, which would prove to be due to sources within or outside of the Milky Way system. On the basis of a suggestion by W. Nernst, Kolhörster concerned himself with this question in 1923 and found that the intensity of the radiation was correlated with sidereal time; this, however, was later disputed by other authors. Substantial progress was not achieved until the further development of atomic physics and radioactivity research. Above all the introduction of the Geiger–Müller counter (1928) was important for the further study of high-altitude rays. With the application of this new instrument Bothe and Kolhörster found that there appeared to be some distinct sources in the sky.

While the opinion had still been generally advocated that these are extremely hard (high energy) gamma rays, the experiments with new equipment unequivocally showed that the cosmic rays are corpuscular in character. It was therefore assumed that there must also be highly accelerated electrons. Yet this assumption also had to be abandoned quickly. In 1929 the Soviet physicist D. V. Skobeltsyn found that the cosmic ray particles could be detected in a Wilson fog chamber. From the deflection of particles in magnetic fields it then was noted that they could not be electrons.

After 1930 an involved study of cosmic rays was initiated. As a result the composition from different elementary particles (mesons, positrons and others) was established as well as the important fact that the particles found in the vicinity of the ground are not identical with those cosmic rays primarily found in the region of space near the Earth; instead they were the result of interactions with the Earth's atmosphere. The detection of cosmic rays on the ground could not be explained on the basis of available knowledge and still today is not fully understood.

After the work of William and John Herschel, as well as Wilhelm Struve, the study of the galaxy made no real advances during the nineteenth century. It is true that several astronomers, among them Proctor, Celoria, Stratonov, and Easton, tried to determinate the distribution of stars in space, but they did not have adequate observational material to work with and did not have the right kind of equipment at their disposal with which better data could be obtained. Not until the beginning of the twentieth century could a solution to the problem be sketched out.

On the basis of observations, however, the investigation of the structure of the galaxy and the attempt to understand the kinematics and dynamics of the system took shape as an extremely complicated problem whose difficulties can only be given a brief outline in the following survey. This problem invariably involved an attempt to determine the distribution of stars in space and the structure of the system as a whole even though we cannot look at it from an outside vantage point.

One attempts to calculate the number of stars per unit volume in every corner of the galaxy (the density function). However, the most obvious means to this end – the determination of the density function from the measurement of the distances of individual objects – had to be avoided because of the great number of stars. Thus the astronomers were necessarily led again to considering the procedures outlined by William Herschel and to work with statistical data concerning the stars. In the first phase of stellar statistics the distribution of stars was assumed to be rotationally symmetric about the Sun, such that it seemed justified to determine the density function with the help of analytical procedures.

In particular, H. von Seeliger, K. Schwarzschild, and C. W. L. Charlier are remembered for their efforts in this area of research. In 1910, building upon the work of von Seeliger, Schwarzschild derived the fundamental equation of stellar statistics, an integral which represents the mathematical relationship between the density distribution, the luminosity function, and that function established from observations which gives the distribution of stars on the basis of apparent magnitude. For the first time the luminosity distribution was considered. It was required because the derivation of absolute magnitudes had shown that the assumption made by Herschel of equal luminosity for all stars was not correct. In connection with statistical investigations the distribution of luminosities is very important information, for in a region of the sky to be studied one sees intrinsically faint nearby stars as well as intrinsically luminous distant stars. Almost all of the nearby stars are counted, but one only counts the distant giant stars. The observed

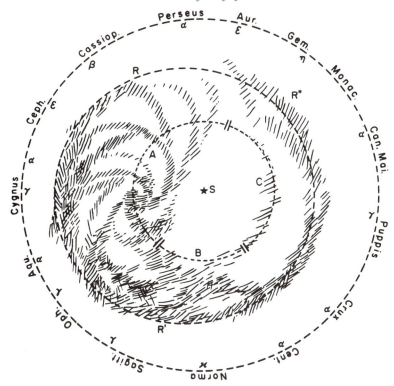

Fig. 42. Schematic diagram of the distribution of stars of our Milky Way system, according to Easton (1900)

star distribution is biased and must be corrected with a knowledge of the luminosity function.

Schwarzschild's analysis gave the following model for our galaxy: a flattened ellipsoid of rotation with the Sun in the center. The density was highest in the center and decreased monotonically in all directions, though this decrease was more pronounced in the direction of the pole than in the direction of the plane of the Milky Way. The dimensions were given as approximately 10 000 parsecs length and 2000 parsecs thickness (oblateness 1:5).

The pioneers of stellar statistics also perceived that this picture of the galaxy was oversimplified. A glance at the starry region of the Milky Way poignantly showed that the distribution of stars was markedly non-uniform. In the same direction in the sky one could see greatly concentrated star clouds next to comparatively empty regions. Finally, it was unlikely that the Sun was really at the center of the galaxy. Star counts in the Milky Way plane also manifested a striking dependence on the galactic longitude, with increased stellar density in the direction of the constellation Sagittarius (galactic longitude $\lambda \approx 325°$) and an especially low stellar density in the opposite direction ($\lambda \approx 145°$). This evidence helped refute the heliocentric picture of our galaxy. However, these facts concerning the apparent distribution of stars were not built into the models of the galaxy. Therefore, several astronomers proposed that methods should be developed which permitted a more flexible consideration of the observed details. Among others, Pannekoek, Malmquist, Seares, and Kapteyn developed such numerical procedures – in essence improvements of the 'star gauges' developed by Herschel. The work of Kapteyn in particular achieved great scientific significance for a while.

Kapteyn began investigating the problem of the structure of the galaxy in 1890. He perceived right away that there was a dire need for more suitable observational material. When he began his work very little information was available: only several dozen stellar parallaxes, a few proper motions for stars fainter than 7th magnitude, and hardly any data for stars fainter than 9th magnitude. However, in order to alleviate the problem as quickly as possible, one observatory was not enough; international collaboration and a highly organized scientific group were required for this. Therefore, Kapteyn developed his Plan of Selected Areas ('Kapteyn gauge fields'). In 1906 he presented this project to the technical world in a comprehensive 82-page publication. According to his suggestion, all available data were to be gathered concerning hundreds of thousands of stars in 206 fields, i.e., positions, luminosities, proper motions, parallaxes, spectral types, and radial velocities. Later 18 fields in the galactic plane were added in order to obtain still more comprehensive data for these important zones. In contrast to the huge surveys of the nineteenth century, the Plan of Selected Areas had a particular scientific purpose.

Kapteyn's plan was generally accepted, and numerous observatories worked on it. The Harvard College Observatory, for example, contributed a photometric survey of stars down to magnitude 16. At Mt Wilson Observatory even 18th magnitude was reached. In Potsdam and Bergedorf spectral surveys to 12th and 13th magnitude, respectively, were carried out. Nevertheless, the completion of the whole project of Kapteyn Selected Areas took several decades.

Once again, with the attempts to map out the structure of the galaxy by means of numerical methods, interstellar absorption was not considered (as was the case before the introduction of the analytical methods). However, this omission, depending historically on various factors, had dire consequences, in particular for the determination of the density function. The apparent magnitudes of stars, which represent a measure of distance, were fainter than would correspond to the true distances, i.e., the derived density function (uncorrected for interstellar absorption) would appear to decrease in all directions. As a result the Kapteyn universe looked very similar to the already-mentioned rotationally symmetric picture. However, because there

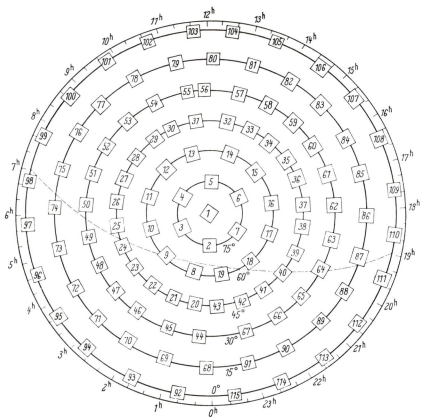

Fig. 43. Arrangement of the Kapteyn Selected Areas Nos. 1 to 115

was no sure evidence for the existence of generally diffused light-absorbing matter, the result was taken to be a reasonably valid description of the galaxy.

It soon became convincingly clear, however, that this model had fundamental deficiencies because a very special group of objects, the so-called globular clusters, allowed one to calculate the dimensions of the whole system.

The globular clusters, rich stellar formations with a strong concentration of stars toward the center, had already been the object of attention for a long time. A particular peculiarity of all globular clusters was their apparent distribution in the sky. One third of all globular clusters are to be found in the constellation Sagittarius, the rest in the same general direction of the sky. This finding gave rise to many possible explanations until the actual spatial distribution of globular clusters was finally understood. However, the most important prerequisite for this was a straightforward procedure for the determination of the distances of individual globular clusters. Such a

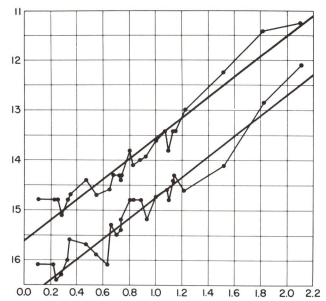

Fig. 44. Period–luminosity relation for δ Cepheids in the Small Magellanic Cloud.
The two curves correspond to the maxima and minima of luminosity, respectively
(according to the data of 1912)

procedure surprisingly came about with the investigation of a group of
variable stars whose light curves resemble those of the prototypes δ Cephei
and RR Lyrae.

In 1908 Henrietta Leavitt published a catalog of 1777 variable stars in the
Small Magellanic Cloud, an irregularly shaped companion of our galaxy
visible in the southern sky. On the basis of the available plates the periods
of sixteen of these stars were determined. These periods ranged from 1.25 to
127 days. Already in this early work Miss Leavitt made the interesting
discovery that the brighter stars have longer periods of light variation. Four
years later she further refined the observational material and clearly
showed that the apparent magnitudes of these stars increase approximately
with the logarithm of the period (Fig. 44). Because the variables in the Small
Magellanic Cloud were situated at approximately the same distance from
the Earth, a relationship was thereby found relating the periods and the
intrinsic luminosities of these stars. Miss Leavitt also proved that there were
variable stars with similar light curves outside the Magellanic Clouds. If it
were possible to measure directly the distances of a few of these stars, then
the exact relationship between period and absolute magnitude could be
derived. As a result this relationship became a method for determining
absolute magnitudes as well as distances. Next, different RR Lyrae and δ
Cephei variables were found in globular clusters. The period–luminosity

relation was apparently the same as that for the Cepheids of the Small Magellanic Cloud. On the other hand the absolute magnitudes of RR Lyrae stars were about $M_v \approx 0$. This was quite accurately known, such that these stars seemed suitable as standard candles.

In 1918 H. Shapley distinguished a calibration of the period–luminosity relation independent of the RR Lyrae stars, in which he calculated the distances of different Cepheids with statistical methods. The absolute magnitudes thus derived for the RR Lyrae stars again came out to be $M_v \approx 0$, such that one could more confidently trust the calibration of the period–luminosity relation.

However, later, when detailed photometric results from globular clusters of the Magellanic Clouds were available, it turned out that the period–luminosity relation derived by Shapley was not universally valid. Moreover, there were noticeable differences with respect to RR Lyrae stars according to whether one considered the Cepheids in spiral arms of extragalactic systems, in the Magellanic Clouds (Population I), or in the globular clusters (Population II). As a result two different period–luminosity relations were needed according to the membership of the Cepheids in the two stellar populations. The most far-reaching consequence of this correction (not achieved until the late 1940s, after W. Baade had established the existence of the two distinct stellar populations) was the doubling of all distances which had been determined by means of Cepheids of Population I.

Yet, independent of these recalibrations, since 1910 the period–luminosity relation has been used to determine the distances of globular clusters.

Shapley carried out such investigations in 1918 and thereby achieved the result which was no less astonishing than the apparent distribution was previously – the globular clusters were symmetrically distributed throughout a huge spherical region, the size of which was hard to imagine. The center of the system of 100 globular clusters lay in the direction of the constellation Sagittarius – the same direction which had already been proposed as the galactic center on the basis of the apparent distribution of stars and the density function. Shapley then assumed that the center of the spherical realm in which the star clusters are situated is identical with the center of the Milky Way system. The whole system which was symmetrically surrounded by, and permeated with, globular clusters accordingly had a diameter of 100 000 parsecs and a thickness of 10 000 parsecs (Fig. 49).

Compared with the results of stellar statistics this was a gigantically large system, much larger than anyone would have ever predicted.

The distribution of globular clusters derived by Shapley was essentially correct. But the dimensions of the whole system required modification. Shapley, too, had not considered the interstellar absorption of light, in accord with the information available. He even felt obliged to explicitly neglect such corrections, for he could not measure any change of color of the starlight in the globular clusters. Later, by means of photoelectric methods published in a separate investigation, Stebbins showed the influence on color of interstellar absorption, such that the distance values had to be improved. The present-day value of the diameter of the galactic halo based on the distribution of globular clusters is 50 kiloparsecs.

According to Shapley's interpretations, vociferously disputed at first, one thing was clear – the whole galaxy was significantly larger than had previously been assumed. The region of space investigated by stellar statistics could at best describe a 'local stellar system'. Whatever structure the whole system really had was admittedly as unclear as before. Here was proven what E. von der Pahlen so aptly described later: 'The impressive edifice created by the old masters lies in ruins and until now it has not been possible to join the emergency shelters together into a solid new structure.'[8]

Though stellar statistics has some inherent drawbacks, the consideration of interstellar absorption led to important results. Since the beginning of the century the evidence had mounted which testified to the existence of generally diffused light-absorbing matter between the stars.

There had been speculations of this kind for quite some time. For example, Olbers tried to resolve the famous paradox of the dark night sky described by him in 1823 by postulating the existence of interstellar material. Wilhelm Struve was of the same opinion; he assumed a light absorption in between 0.1 and 3.8 mag/kpc as a consequence of the light-absorbing material.

However, the modern procedures of astrophysics gave us the first concrete evidence, especially the use of photography since its application by Max Wolf and E. E. Barnard. Photography revealed the markedly non-uniform distribution of stars in the region of the Milky Way, which was then thought to have extensive regions empty of stars. But it was shown that the unequal distribution of stars in these regions is only apparent and made to seem so due to giant clouds of dust placed between the stars. Using the star-count data as a function of apparent magnitude, Wolf and others developed methods which rendered possible the derivation of distances, dimensions, and the absorbing effect of clouds from the observed deviations of star counts from the expected values (Fig. 45).

There was still more evidence implying the existence of light-absorbing matter. One of the most impressive results was the discovery of a 'zone of avoidance' along the galactic equator. In an irregularly formed band along the Milky Way no galaxies are seen, whereas these objects are distributed quite uniformly over the rest of the sky. As a result one might conclude that the nebulae really had such an orientation with respect to the plane of the Milky Way. However, as it turned out, the nebulae are really extragalactic systems, so this interpretation seemed unlikely. Instead, the apparent distribution of nebulae demonstrated that there was a layer of material coinciding with the plane of symmetry of the Milky Way system.

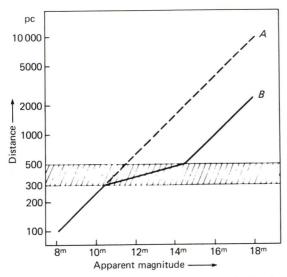

Fig. 45. Schematic diagram of absorption through a dark cloud. The distances estimated on the basis of star counts per square degree systematically deviate from those in an unbiased field because of the absorption (A – unbiased field, B – field affected by absorption)

This interpretation was greatly substantiated by the observation of other stellar systems, for example, the galaxy NGC 891, which clearly exhibits a layer of non-luminous material in the region of its galactic equator (Fig. 46).

Eventually many phenomena also testified that the interstellar material was not only concentrated in clouds in the vicinity of the galactic equator, but it was quite generally diffused throughout the galaxy.

In 1904 J. Hartmann investigated the radial velocities of the spectroscopic double star δ Orionis. One of the spectral lines showed a remarkable behavior. The line of ionized calcium did not partake of the periodic line shifts caused by the orbital motion. Therefore, Hartmann concluded that this 'stationary calcium line' probably did not arise in the atmosphere of the star; rather, it is caused by interstellar calcium gas. Slipher and others were able to show later that the 'stationary calcium line' is observed in the spectra of many other stars. Besides, numerous other stationary lines were gradually discovered and were attributed to other elements. This was further proof for the existence of interstellar matter.

R. J. Trumpler, who began making observations of open star clusters in 1930, provided the crucial evidence. He was able to show that certain clusters have approximately the same linear size, so that the measurement of their apparent diameters permitted the determination of the distances. Then with increasing distance one can expect a predictable decrease of

Fig. 46. The extragalactic stellar system NGC 891, photographed by J. E. Keeler at
Lick Observatory in 1899. The orientation of the object with respect to the direction
of the terrestrial observer lets one clearly view the zone of absorbing material in the
galactic plane

luminosity for the stars in these clusters; Trumpler found that the 'photometric distances' were consistently greater than the 'geometric distances', implying that the light intensity decreased more sharply than the square of the distance. From this he likewise deduced the existence of generally diffused interstellar matter. On the basis of all these discoveries a new important area of modern research was inaugurated – the study of the properties of this interstellar matter.

The interstellar material was of great significance for the further development of stellar statistics. Although on the basis of scanty information one assumed a uniform distribution and thus made calculations with an assumed universally applicable absorption constant, there appeared some evidence which had not been uncovered in the investigations of the absorption in different directions. Indeed, a decreasing star density in all directions from the Sun was again noted, but for distances greater than 1000 parsecs an increase of density was observed. This especially proved true in the already-mentioned direction of $\lambda \approx 325°$. This finding was consistent with the picture that the Sun finds itself in a region of the galaxy of relatively high star density, which, however, represents a small piece of the whole system. Thus it must appear very likely that the great amount of space between the globular star clusters outside the 'local system' is likewise not empty of stars. The closer study of this question could at best succeed only with the brightest objects, for only these could still be detected at great distances. Moreover, one also has to determine the distances of these objects. Their study then led to 'a very patched-together configuration whose intervening space we can think of as filled up, on the basis of observations in the vicinity of the Sun'.[9]

One of the most substantial results of these investigations was the determination of a density maximum which again coincided with the center of symmetry of the globular cluster system. If one considers the intrinsically very bright objects to be indicators of the existence of other stars too, this implies a very massive center for our galaxy.

Now the results of kinematic investigations were tied together with these findings, from which a consistent picture could on the whole be derived. This at the same time led to the development of a dynamical theory of the Milky Way system.

The investigation of the kinematics of the galaxy rests in part on the methods of classical positional astronomy and partially on the use of the interwoven methods of astrophysics.

The first hint that the stars do not have absolutely fixed positions in the sky came from Halley. In 1718, by comparing the observations from the star catalog of Ptolemy with contemporary positional data, he determined that the stars Sirius, Aldebaran, and Arcturus had changed their positions. This interesting fact, which as a result implied that the constellations changed their shapes over the course of millenia, gave cause for very precise positional measurements and led to certain proof of the existence of proper motions by T. Mayer in 1760. The laws of stellar proper motions were first put together on the basis of his extensive material.

Mayer assumed that the stellar proper motions are randomly distributed in direction and magnitude. They represent a component of the peculiar motion of the stars. Then a peculiar motion of the Sun – to be considered as very likely – must be reflected in a systematic distribution of proper motions of other stars, from which the apex of solar motion in the sky could be determined. William Herschel first determined the direction of the apex in 1783, if only very crudely. Apex determinations were later derived and refined from improved proper motion data. Finally in 1885 they were supplemented through the use of radial velocity measurements, which permitted a determination of the velocity of solar motion* (Fig. 47).

The accurate study of the elements of solar motion led in 1915 to an extraordinarily important finding which later became the basis of the dynamical theory of the Milky Way system. With the investigation of the radial velocities of B stars it was shown – after the influence of solar motion was removed – that these stars exhibited a variation of velocity with galactic longitude. Freundlich, von der Pahlen, and Oort, in particular, more accurately studied this effect in the years 1923 to 1928, basing their work on that of Gyllenberg. The dependence of radial velocities on galactic longitude manifests itself in the form of a double sine wave (Fig. 48) in a graph, with the maxima and minima at $\lambda = 10°$ and $\lambda = 190°$, and $\lambda = 100°$ and $\lambda = 280°$, respectively. For the longitudes 325° and 145° the peculiar radial velocities average out to zero. Here for the first time the already-known directions toward the center and anti-center of the system were reflected by kinematic investigations. Moreover, a study of the double waves for objects at different distances showed that the amplitude of the sine wave becomes greater for more distant objects. Also, the effect was not only limited to B stars.

It could be concluded that these kinematic data were of considerable importance for the understanding of the Milky Way system. Lindblad and Oort also based their dynamical theory of the Milky Way (1926/27) on such data. Oort attempted to unite the kinematic facts discussed and the results concerning the mass distribution (the flattening of the ellipsoid of rotation with mass concentration in the center). He concluded that the whole system rotated in the Milky Way plane about an axis perpendicular to the direction of the center. If this is the case, Oort concluded, then the proper motions must also exhibit a functional dependence on galactic longitude which likewise has the form of a double wave but which is displaced from the curve of radial velocities by a shift of 45°. Although making these observations involved numerous difficulties, Oort succeeded in showing such a double

---

* One can also determine the velocity of solar motion if the distances of the stars of measured proper motion are known. (Tr. note)

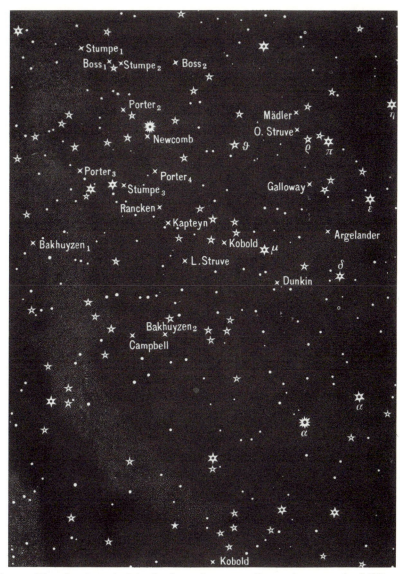

Fig. 47. Direction (apex) of the Sun's motion, according to the research of different authors

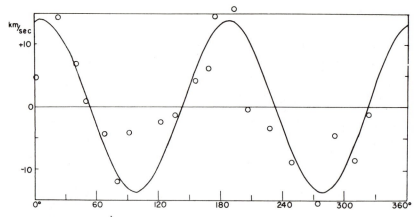

Fig. 48. Double wave of radial velocities as a function of galactic longitude –
evidence for the rotation of the galaxy

wave, such that a rotation of the whole system seemed confirmed. Accordingly, the individual stars must move about the center of the Milky Way in approximately Keplerian orbits. From the position and the velocity of the Sun the total mass of the Milky Way system could be shown in this way to be approximately 10^{11} solar masses.

The study of the details of the dynamics was just as hampered as the investigations of the structure of the galaxy. Both, moreover, represent today two of the most difficult problems of astronomy. For many findings obtained today a completely new observational procedure must be used, like, for example, radio astronomy for the mapping out of the spiral structure of the Milky Way.

Many results concerning the Milky Way system are based on the astronomical study of other stellar systems in the universe, which have the advantage, in contrast to our own stellar system, of being observable from a distance as a whole. Indeed, the reality of distant stellar systems in the realms of space could not be confirmed until relatively recently. Even in the first decades of the twentieth century it was very uncertain whether cosmic objects generally existed outside the galaxy or whether all these objects were constituents of our own stellar system.

Huggins believed that the Andromeda Nebula was a planetary system in a state of evolution. In 1890 Agnes M. Clerke quite decidedly argued that there are nebulae outside the Milky Way system. Although Scheiner unequivocally proved that the spectra of different nebulae are identical to those of regular stars, this did not signify a breakthrough concerning the establishment of the extragalactic nature of the objects.

As for the special nature of the Andromeda Nebula, even in 1907 evidence seemed to testify to the notion that it is not at all, like many believed, outside the Milky Way system. Bohlin published a value for the parallax of this

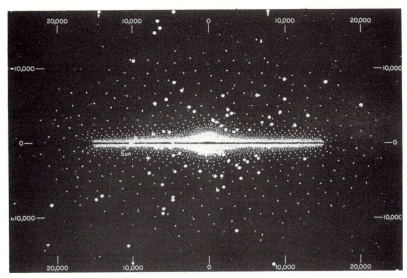

Fig. 49. Model of the Milky Way system by J. S. Plaskett (1939)

object, from which a distance of approximately 20 parsecs results – an embarrassing mistake, for this value is on the order of 100 000 times too small.

Applicable to this were H. D. Curtis' results of 1917, which were based on the discovery of a nova in the Andromeda Nebula. Curtis assumed that the absolute magnitude of the nova at outburst maximum agreed with the values for other novae known to be relatively nearby, and from the difference of apparent and absolute magnitude, without a consideration of interstellar absorption, he derived a value for the distance to the nebula. He found a distance of one million parsecs, which indicated that the nebula was far outside our galaxy. However, the result was not generally accepted, for there was no certain proof that the nova belonged to the Andromeda Nebula.

A scientific solution to this extremely important question for the understanding of the structure of the cosmos only came about with the purposeful application of new information and through the employment of previously unused technical means. In 1923, with the use of the new 2.5-m reflector at Mt Wilson Observatory, Hubble succeeded in resolving the outer regions of the Andromeda Nebula (M 31) and the Triangulum Nebula (M 33) into stars and also succeeded in discovering among these individual stars variables of the δ Cephei type. Under the assumption that these Cepheids obey the same period-luminosity relation as that of the other objects of this type so far discovered, Hubble determined the absolute magnitudes from the periods and then the distances of the objects from the differences between the absolute magnitudes and the apparent magnitudes. Although the values then determined were too small by 50 per cent because of the incorrect calibration of the period-luminosity relation (the value used today from M 31 is 2.2 million light-years), there was no longer any doubt that this type of 'nebula' is situated far outside the Milky Way system in space. Complete certainty concerning this question was achieved within a few years when Hubble resolved the outer regions of a total of about 125 'nebulae' into stars and in several objects identified numerous Cepheids, whose study led to even greater distance values than that for the Andromeda Nebula.

Once again, with the detailed study of the extragalactic stellar systems, a new extensive portion of study of the universe was introduced. Astronomy expanded into still greater realms of space, of which hardly anyone at the time of Herschel could have had any conception.

RELATIVITY THEORY AND ASTRONOMY

The close interdependence of twentieth century astronomy and physics is not limited to the importance of quantum mechanics for stellar evolution. Albert Einstein's General Theory of Relativity is another area of physics which has developed and which still exerts a strong influence on the study and interpretation of the cosmos. This pioneering theory signified a revolution in the existing models of space and time and eventually created the theoretical bases for the investigation of the cosmos as a whole.

The General Theory of Relativity is the theory of gravitation. The theory says that the confusing Newtonian notion of action at a distance is no longer necessary. Newton himself and some of his successors were expressly dissatisfied with the idea that forces emanating from masses must work with infinite speed, instantaneously through empty space to the deepest realms of the cosmos. Einstein succeeded in eliminating the need for speaking of forces acting at a distance, as the research of Faraday, Maxwell, and H. Hertz had shown that forces resulting from electromagnetic effects were not forces acting at a distance. A magnet or an electric charge does not make its effect on another magnet or electric charge with infinite speed; instead, we speak of magnetic or electric fields carrying the energy through the intervening space.

Though Einstein arrived at a field theory of gravitation, the connection between the gravitational field and space is of a different nature from an electromagnetic field. The masses warp the space, which is to be understood as a four-dimensional space–time continuum, namely a particular structure, and the motion of masses inversely determines this.* The structure of space is in general non-Euclidean; it is a warped (Riemannian) space. The planets move, for example, due to their inertia, according to the structure of space influenced by the Sun. Relativity theory assumes that the physical

* Einstein said: 'Space tells matter how to move, and matter tells space how to curve.' (Tr. note)

structure of the gravitational field is identical with the local geometric (metric) structure of the space–time world.

However, the astronomers, especially during the whole nineteenth century, piled triumph upon triumph for the Newtonian theory. To what degree were these proven foundations of classical theory brought into question by relativity theory? The Einsteinian theory did not unequivocally invalidate all of Newtonian mechanics; rather, the new theory showed that Newtonian physics can be considered a limiting case of the Einsteinian theory. For small masses and relative velocities which are small compared to the speed of light the relativistic equations reduce to those of classical physics. If, however, these conditions are not fulfilled, there arise consequences which are not inherent in Newtonian physics and which are completely incomprehensible in its scope. The Einsteinian relativity theory thus laid out quite general and comprehensive statements, as did classical mechanics, and also introduced concrete improvements to Newtonian celestial mechanics.

Understandably, if relativity theory were to be accepted, its effects must be somehow observed in nature. Here lies an important connection between relativity theory and astronomy.

In 1911, when his theory was still being put together, Einstein was discussing the possibilities of confirming the theory with the young Potsdam astrophysicist E. Freundlich. As a consequence three relativistic effects became recognized which could be used for confirming or refuting the theory.

The first was a peculiarity in the orbital motion of the planet Mercury about the Sun. Newtonian mechanics predicts that the orientation of the major axis of the elliptical orbit should change. But the observed value of this 'advance of the perihelion' differed by 43 arc seconds per century from the Newtonian value.

This effect had already been known to astronomers for more than fifty years, but it called forth a great variety of explanations. In 1859 U. J. J. Leverrier had noted in a study of Mercury's motion that a systematic difference was showing up between the Newtonian theory and the actual motion which corresponded to an advance of the perihelion of about 41 arc seconds per century. In 1895, within the scope of celestial mechanical investigations, the American astronomer S. Newcomb reviewed all the available observational material – data for 200 years, during which time the planet had made 850 revolutions about the Sun. Newcomb also determined that the perihelion of the planet's orbit had advanced 5600 arc seconds (1.56 degrees) per century, whereby a remainder of 43 arc seconds per century could not be explained by Newtonian physics. A well-known attempt to explain this anomaly was the hypothesis of the existence of a planet revolving about the Sun within the orbit of Mercury. However, this planet (Vulcan) was not found.

The General Theory of Relativity not only explained this perihelion advance, but directly predicted the value. Thus the theory was confirmed in the realm of accurate measurements and it simultaneously solved an old riddle. As might be expected, similar perturbing effects were predicted for the other planets; they are, however, substantially smaller, as the effect depends on the distance from the Sun and therefore becomes more difficult to demonstrate clearly. For example, the relativistic perihelion advance for Venus amounts to only 8.6 arc seconds per century, and for Jupiter it is only 0.06.

A second prediction, which had not been borne out by existing observations, concerned the bending of light. According to relativity theory a light ray moves in a straight line, but if it passes a massive object (i.e., travels through a region of warped space), it will change its direction. If this proved to be true, then the position of the star whose light passes by the solar limb will appear to exhibit a slight displacement. Now the Sun describes an apparent orbit in the sky due to the Earth's actual orbit about the Sun. This carries the Sun through the whole zodiac in the course of a year. There will always be some stars in the same direction of the sky as the Sun; these stars will be found several months later in the night sky. The deflection of light in the gravitational field of the Sun must therefore be quite suitable for examining this bold assertion of relativity theory. However, because the stars cannot be seen in the daytime sky, it was proposed to obtain the star positions with the greatest possible precision in the direction of the Sun during a total solar eclipse and then later to measure them again in the absence of the Sun and to compare the two sets of positional values. It was clear that this was an extraordinarily complicated technical task. One had to account for a great number of possible errors of measurement; besides, the value predicted by the theory was small – for a star right at the solar limb the theory predicted a displacement of only 1.75 arc seconds.

In the summer of 1914 a German research expedition under the direction of Freundlich travelled to Theodosia in the Crimea in order to carry out the important measurements. This undertaking was thwarted by Germany's declaration of war on Russia (1 August 1914). The German scientists were detained for a time, then expelled from Russia. Not until five years later did the English take up this problem again. Eddington, Crommelin, and Davidson outfitted two expeditions which made observations on 29 May 1919 in Sobral (Brazil) and on Principe Island (West Africa).

The English scientists succeeded with the difficult undertaking. Their value of 1.98 arc seconds for stars at the solar limb, derived from numerous photographs, was generally regarded as a satisfactory confirmation. Later measurements resulted in somewhat different values, but all testified to the reality of the bending of light. The proof of this tiny effect, which cannot be explained within the scope of classical theory, made relativity theory famous overnight. However, the technical details of the theory were understood by only a select few.* Nevertheless, the unusual amount of public interest did have an extremely wide-ranging effect. It seems to have caused the German government to set aside funding for the construction of the Einstein Tower for Solar Physics in Potsdam, because the government feared that research in this pioneering area founded by a German might otherwise be dominated by foreigners.

Einstein predicted a third relativistic effect – the red shift of spectral lines in a gravitational field. An atom which radiates in a gravitational field emits electromagnetic waves of lower frequency than the same atom outside the gravitational field – the wavelengths of the emitted radiation are shifted toward the red.

Next it was attempted to confirm the effect in the gravitational field of the Sun. Success was expected because of the great accuracy with which one could determine the wavelengths of lines in the solar spectrum. Difficulties ensued, however. Although the expected line shifts were within the range of measurability, these shifts were masked by many other considerable effects. There were widely varying opinions concerning the measured value, and the problem of relativistic red shifts remained a major task for astrophysics. Not until recently was the unscrambling of the different effects accomplished and the value predicted by relativity theory confirmed for the limb regions of the Sun.

Investigations which involved the measurements of relativistic red shifts in the spectra of small massive stars met with other difficulties; for, in order to determine the theoretical value, very precise information concerning stellar masses and radii is required. Indeed, the red shifts in the spectra of different white dwarfs had been found, but the accuracy of the values was not sufficient to speak of quantitative confirmation. Then in 1960 it was indeed accomplished in a study of the red shift by astrophysical procedures, for the measurement of the shift was also made possible by the consequences of the Mössbauer effect in the gravitational field of the Earth. The agreement between theory and measurement was thereby achieved.

RELATIVITY THEORY AND COSMOLOGY

The ideas developed by Einstein were predestined by their nature to make possible statements concerning the universe as a whole, for as a theory of the gravitational field, relativity theory concerned itself with the single universal interaction known to us. Such age-old questions as the structure of the cosmos and whether it is finite or infinite were therefore placed by this theory in completely new perspectives. All modern cosmological investigations rely on the General Theory of Relativity.

* The story goes that a reporter once asked Eddington if it were true that he was one of only three people in the world who understood Einstein's theory. Eddington's silence prompted the reporter to commend him on his modesty. But Eddington said that it was nothing of the kind – he was just trying to think of who the third person might be! (Tr. note)

Fig. 50. Albert Einstein

Soon after the publication of his theory Einstein himself attempted to derive a model of the cosmos as a whole. In addition to his modified relativity theory he used the fundamental cosmological principle, which states that the observed universe would look the same to an observer in any part of the universe (isotropy), and that the average density of matter is everywhere the same (homogeneity). Einstein's result was that the universe as a whole represents a static spherical space, in accordance with Riemannian geometry. The sum of all spatial distortions resulting from cosmic masses in practice describes a warping of the space–time continuum as a whole, and the cosmos represents a closed and unbounded, but not infinite, space. Like the surface of a sphere (the two-dimensional analogy) is finite but boundless, so is the volume of this spherical space also boundless. A light ray which originates somewhere in this non-Euclidean cosmos will return to its point of origin after a certain amount of time.

This model of the cosmos developed by Einstein was a universe without change. It had a defined radius which could be calculated from certain information, and this never changed. In 1922 the well-known Soviet mathematician and physicist A. A. Friedmann published another self-sufficient solution to the relativistic cosmological equations. In his work 'Concerning the bending of space', which indeed largely remained unknown, he showed that the assumption of a static, non-evolving cosmos is not at all reconcilable with the General Theory of Relativity. Instead, the cosmos as a whole must either expand or contract. Following Friedmann, and in part independent from him, H. P. Robertson, Eddington, Otto Heckmann, and other scholars obtained the same result. This enables one to predict that the spectral lines of distant stellar systems should exhibit systematic red or blue shifts. This seemed absurd to most physicists. While relativity theory was still trying to gain a toehold in established science, the prediction that the universe was expanding or shrinking was considered an example of the inapplicability of the theory to cosmological questions.

However, it was only a few years later that experimental astrophysics provided one of the most exciting scientific discoveries of the century – the recession of the spiral nebulae.

This significant result was the fruit of a nearly twenty-year-long research project in this area. It was thus begun long before the General Theory of Relativity had provided its cosmological consequences. Even more, the first spectra of nebulae which were suitable for a demonstration of the Doppler shifts of absorption lines were already available before the extragalactic nature of these objects was proven. V. M. Slipher of Lowell Observatory obtained such data for the first time in 1912. By the year 1925 he had obtained radial velocity measurements of more than fifty objects which exhibited very large velocities, up to 1100 km/s. Already at that time there was evidence that the magnitude of the line shifts depended upon the distance. However, in order to substantiate this suspicion, they needed

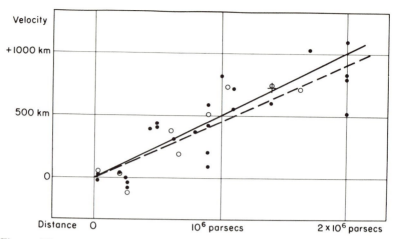

Fig. 51. The expansion of the extragalactic stellar systems according to Hubble

measurements of faint (i.e., distant) nebulae, as well as precise information concerning the distances themselves. For this only one instrument was good enough – the new 100-inch Hooker reflector of Mt Wilson Observatory, which had already helped demonstrate the extragalactic nature of the Andromeda Nebula. Due to the new kind of camera optics of this instrument's spectrograph, spectra of nebulae were obtained with a dispersion never before achieved. Hubble and Humason made use of the longest possible exposure times for 65 extragalactic objects over a great range of distance. In 1929 they showed that the radial velocities of the nebulae were directly proportional to their distances (Fig. 51); furthermore, except for a few nearby nebulae, all had red-shifted spectral lines. This sensational finding directly indicated that the universe was expanding and that the main prediction of Friedmann's model was confirmed.

The substantiation of the cosmological models resulting from the theory of relativity had important repercussions for the subsequent development of modern cosmology. It by no means freely signified an end point; quite the contrary – it was the beginning of intensive theoretical and experimental astrophysical activity which continues undiminished. For example, though Friedmann's evolutionary cosmos was confirmed by means of the discovery of the red shifts in the spectra of galaxies, it could still not be decided whether the universe was finite or infinite. This can only be decided by means of even more extensive astronomical observations than are presently possible. So far the observations have not allowed us to determine if the average warping of space is positive, negative, or zero. Cosmological research inspired by Einstein's theory of relativity has many fundamental problems to solve and has become a central concern of many research groups in the past few years.

4

Technology and the organization of research

Since the invention of the telescope in 1609 astronomical observational instruments have been one of the important factors for the progress of astronomy. Every breakthrough which gives us greater understanding of the nature of cosmic processes and all achievements of astronomy depend on progress in the technological aspects of observational detectors and measuring instruments. The construction of astronomical measuring devices and observational instruments places great demands on the different areas of technology and science, in particular on optics and precision mechanics. The development of these technical disciplines is closely tied to the development of the means of production. In those phases of astronomy which engendered fundamental science, the technical research problems were, therefore, very closely connected with the state of economics and production as determined by society.

The eighteenth and nineteenth centuries, characterized by the further growth of the bourgeoisie and the increased use of machinery, also made possible immense progress in the area of construction of astronomical research instruments. Precise mechanical finishing procedures, material adaptation methods, and production methods were substantially stimulated by the origin of machine production. These very improvements, which were useful for the production of steam engines or looms, also were applied to the production of astronomical instruments. On the other hand, one should not overlook the highly specific demands which had to be placed on the astronomical instruments. They led to a further development of a great variety of different techniques, which then were also applicable to other areas. Among other things, the interconnection of economics, technology, and the construction of astronomical instruments points to the fact that at first the best instrument makers worked in The Netherlands, England, and France. Not until the nineteenth century did German craftsmen also succeed in playing a leading role in this area, whereby a close collaboration with the English and French manufacturers was an important criterion for the rapid introduction and subsequent development of precision instrument construction. This was the case in general. A. Hamann, the first German who produced factory-made lathes, had for five years personally studied the experience of the English on site.

In earlier phases of astronomy's development the astronomers were for the most part also the instrument builders. With the great interest in clocks and sextants – above all for shipping – a change was initiated in 1700 which led to a division of labor. This change led to the origin of a new vocation, the production of precise scientific instruments.

The active industrial production of astronomical instruments grew throughout the course of the nineteenth century. This was closely connected with the development of measuring instruments, tools, and technical equipment of other sciences. For example, one also needed the photometer for the irregularly developing lighting industry, and the spectrometer for spectral analysis in chemistry. The producer of astronomical instruments in general made a wide variety of products – glasses, microscopes, micrometers, drawing tools, scales, and compasses.

Concerning the development of astronomical instruments there were above all three goals. The first concerned the creation of ever more powerful telescopes and with it the observation of celestial bodies at greater and greater distances. The second goal was to make possible ever more precise positional determinations for celestial mechanical investigations. Thirdly, one wanted to obtain ever more detailed information concerning the light of the stars. This was a highly important program for science and technology. It called for larger optical equipment, improved mountings, a perfected theory of instruments, and all astronomical information needed for the reduction of the observations, as well as a close collaboration between astronomers and instrument makers. The result was a series of different instrument types, including custom-made auxiliary instruments.

Primarily, there are two basic types of astronomical telescopes which arose almost simultaneously: telescopes with a convex lens as the objective (the refractor; since 1609) and telescopes with a concave mirror as the objective (the reflector; since 1616). Ever since these two instruments came on to the historical scene they have been in a 'contest' with each other. Sometimes the refractor seemed to be the better instrument for research. Sometimes the reflector proved more applicable. Not until the twentieth century did the reflecting telescope emerge overall as the more widely used type of instrument for the critical research needs of astronomy.

At the start the reflector had the advantage over the refractor because it exhibits no chromatic aberration. The first breakthrough for refractors was therefore the invention of the achromatic lens by Dollond (1758). Nevertheless, there was one major drawback with the achromatic telescopes – it was not possible to construct large crown and flint disks of sufficient quality. They were limited to an objective diameter of about 10 to 12 cm. This was the main reason why the greatest observers at the end of the eighteenth century worked with reflectors; the diameter of the mirror could be as much as ten times larger than the largest possible lens diameter. As a result an enormous capability for range and resolution was achieved.

The great master of reflecting telescope production in the last quarter of the eighteenth century was William Herschel. His instruments were a decisive reason for his success. Herschel ground and polished an incredible number of mirrors. For example, according to his own records, he made more than 400 mirrors in six to seven years. Of every batch he selected the best and then compared these with the best of other batches. There were no scientific testing methods for the mirrors, so he had to use this empirical but foolproof method of selection to obtain good quality mirrors.

Herschel's numerous small telescopes were in very great demand and widely distributed. For example, in Germany J. H. Schroeter had a number of telescopes with mirrors by Herschel. Other, larger instruments by Herschel were to be found in St Petersburg, Madrid, and Göttingen.

The largest mirrors made by Herschel had diameters of 50 and 122 cm. They represented the absolute high points of achievement of optics at that time. The larger of the two instruments, the 40-foot telescope, achieved legendary fame (Fig. 52). This dynamic, huge construction, with a gigantic scaffolding for the mounting and motion of the telescope, had a total mass of 30 000 kg. The cylindrical tube, greater than 12 m in length, was put together with single plates. The technology used for the formation and assembly of these metal plates was the same as that used in the production of bridge arches and smokestacks. At one point Herschel engaged more than forty workers at the same time for the construction of the giant telescope.

Herschel's endeavors concerning the area of telescope construction were no less pioneering than his scientific research. The methods worked out by him were adopted and further developed by his contemporaries, J. Ramage and N. S. Carrochez, among others, but above all by W. Lassell and W. Parsons, Earl of Rosse, a rich Irish nobleman. Through experiment and comparison of mirrors which were produced by means of different procedures, the last-mentioned hit upon a number of rules according to which mirrors of good quality could be made. Lord Rosse also systematically experimented with the alloying of metal mirrors. He designed and produced 'patchwork' mirrors, including a 90-cm mirror made of sixteen fused-together metal plates.

The most famous item by Lord Rosse was a massive mirror 1.8 m in diameter, which he finished in 1842. The mass of the metal mirror amounted to 4000 kg, such that a carefully designed mechanical mounting for the reflector was required. In this regard he essentially surpassed Herschel. The giant telescope of Lord Rosse, known as the Leviathan of Parsonstown, was a nineteenth century master-work of astronomical observational technology. The procedures developed by Lord Rosse in connec-

Fig. 52. The remains of the large Herschelian reflector in the garden of the former
Royal Observatory, Greenwich (1970)

tion with the construction of his telescope represented an important
contribution to the methods of producing larger reflectors; these larger
telescopes were later built by Lassell and Grubb, who laid the foundations
for the establishment of the modern reflectors of the twentieth century.

Meanwhile, at the turn of the nineteenth century a new epoch of
refractors was initiated, because it became possible to produce larger crown
and flint lenses. The rapid development in this area had been substantially
hindered by the establishment of an English monopoly which lasted for
quite some time. France and Germany wanted to be able to produce their
own optical glass. This was also necessary because it was difficult to obtain
this material from England, owing to the political situation concerning
England and France at that time.

In 1773 the Paris Academy of Sciences established a prize for the
manufacture of crown and flint glass, without which the desired progress
would not have continued. This was less a consequence of failing efforts

than of the general state of the glass industry; for specific technical reasons flint glass for astronomical objectives can only be produced in factories which make a considerable quantity of this kind of glass. The first successful flint glass producer at the beginning of the nineteenth century was Dufougeray, owner of a crystal factory in du Creuzot. D'Artigues and Cauchoix, respectively, produced greater quantities.

In Germany J. von Utzschneider resolved to produce astronomical objectives. He became a partner in the precision-mechanical–optical works founded in 1802 by Reichenbach and Liebherr. Fraunhofer later worked there. The collaborative work of these men characterized the broad spectrum of problems which had been associated with the production of useful refractors. Utzschneider was the type of fundamental capitalist entrepreneur with business sense and organizational talent; Reichenbach understood how to readily apply scientific knowledge to technology; and Fraunhofer was the genial theoretical and practical optician. It is noteworthy that Reichenbach had acquired his knowledge in machine construction, metallurgy, and instrumentation as a former pupil in an English military academy in the 'honoured land of technology'.

Utzschneider established a glassworks in Benediktbeuren in Bavaria. The firm was principally run by P. Guinand, who had already gained experience in Switzerland with the casting of larger glass pieces. After Reichenbach had founded the Mathematical–Mechanical Institute in Munich on the basis of his first endeavors, he received significant funding as well from the Bavarian Academy of Sciences and the newly-founded Topographical Bureau. The official commission in charge of the drawing-up of a military–geographical map of Bavaria had expressed concern with regard to the lack of instruments of adequate quality.

Bavaria's marked interest in military–scientific undertakings becomes understandable if one considers that it was one of the first German states to take part in the war of intervention against the revolutionary French Republic.

Pioneering work was accomplished in the German workshops. Fraunhofer worked on the theory of achromatic objectives, developed methods for a more accurate determination of the index of refraction of glass, and created innovative techniques of glass production. For mechanical construction of telescope mountings Reichenbach considered expansion coefficients and the flexure of metals; he facilitated the operation of telescopes by more accurate calibration and reduction of weight. He contributed substantially to the increase of measurement accuracy. Graduated machines were produced by him which made possible a precision of setting only previously achieved by Ramsden and Gambey in England and France. In particular, Repsold further improved the precision of graduated divisions so that more accurate measurements of celestial bodies could be made, and the mechanic Oertling of Berlin then came out with automatic ruling machines.

Fig. 53. J. Fraunhofer

The workshops in Munich were the first large-scale factories specifically for astronomical instruments. In the first third of the nineteenth century they supplied practically all observatories of note. The production program was extremely varied and included microscopes, mathematical tools, theodolites, refractors of large dimensions, quatorials, meridian circles, and transit instruments. All products of these workshops were highly regarded. The prices were high and delivery was slow.

The development of instruments proceeded hand in hand with the uses of the instruments. The correspondence between Fraunhofer and the astronomers Struve, Bessel, and others bears witness to this. The whole history of Fraunhofer's important instrument factory is a proof of the inseparable connection of science, technology, and production on the one hand, and scientific knowledge and requirements on the other.

The largest instrument finished by Fraunhofer was the refractor for the Russian observatory in Dorpat, which the astronomer Wilhelm Struve had ordered. The refractor, in all regards a masterpiece of its time, had an objective diameter of 24.4 cm and a focal length of 4.133 m. The fact that one could read the *Journal de Paris* at a distance of 250 m with the objective of the Dorpat refractor excited the greatest sensation among the public. Struve's pioneering work in the area of double star research to a great degree depended on this excellent instrument.

The heliometer which Bessel ordered in Munich proved to be similar. In 1829 it was delivered by Utzschneider to Königsberg. A micrometer screw allowed the precise setting of the objective head and the setting of the measurable shift of both the objective halves. For example, the first stellar parallaxes measured with this instrument by Bessel corresponded to a mechanical shift of the two objective halves of barely 0.005 mm.

The work of Fraunhofer, above all, initiated a new epoch of refractors. The Munich-based monopoly in this area lasted for a long time after Fraunhofer's death, where the firm was carried on by Merz with the mechanic Mahler. The large refractor for the Russian observatory at Pulkovo, with a diameter of 38 cm, was a famous telescope by Merz and Mahler. Orders for refractors of similar dimensions came in from observatories all over the world. American astronomers also established their observatories with European instrumental masterpieces. Thus the newly-founded observatory of Harvard College ordered a refractor from Merz which was similar to the Pulkovo instrument. Merz and Mahler succeeded in producing lenses of nearly 50 cm in size. However, the firm under the direction of Merz ultimately based its reputation on the successes of Utzschneider, Reichenbach, and Fraunhofer, and thus gradually lost the connection with development.

France, England, and eventually the USA developed their own capabilities for making optics and instrumentation. In the USA the opticians A. Clark and A. G. Clark succeeded in creating objectives whose dimensions have still not been surpassed. A. Clark created a 30-inch diameter objective for a new refractor for Pulkovo Observatory (Fig. 54), and in 1897 his son A. G. Clark produced the largest objective yet made for the Yerkes Observatory. It has a diameter of 40 inches (Fig. 55).

The Clarks were dyed-in-the-wool empiricists; the quality control of their lenses was ensured by them on site at observatories. They improved the objectives until they were satisfied with the quality of the images. Exact tests of the Yerkes objective carried out later showed that it was one of the best lenses ever made.

Fig. 54. The large refractor of Pulkovo Observatory

Fig. 55. The 40-inch refractor of Yerkes Observatory

Fig. 56. Mounting of Alvan Clark & Sons, for the testing of the large objective for
the Pulkovo refractor

Eventually in Germany there was another manufacturer of first-class optical instruments; we refer to the firm of Carl Zeiss in Jena, which was fully steeped in the Munich tradition.

Carl Zeiss was an optician and mechanic; Ernst Abbe joined as a theoretician; and O. F. Schott had great experience in the production of glass as a chemist. The success of the Zeiss workshop, which rose to become an international firm for optical production, showed once again that the interconnected problems of complicated production can only be solved by means of precision mechanics, glass technology, and optical theory.

Zeiss' optical workshop in Jena was already fifty years old when Abbe organized a department for the construction of astronomical instruments. The department quickly developed into a successful, internationally respected business. At the beginning of the twentieth century they could produce large optics – up to 65 cm in size. In addition to astronomical optics, mountings for large instruments also became part of the production line. The astronomical and mechanical skills at Zeiss were directly put to the test with the building of large refractors like the 65-cm refractor of the Babelsberg Observatory, which placed considerable demands on the mounting with its 11-m focal length. F. Meyer in particular was instrumental in the development of astromechanics at the firm of Zeiss. Among other things he developed the relief mounting, used worldwide, in which the mechanical forces of the considerable mass of a large telescope are counterbalanced by means of a separate relief system.

Meanwhile, at the turn of the century the dimensions of the large lenses reached a limit not yet exceeded. In the area of the production of large reflectors, several improvements of basic significance had already been achieved. A new epoch of reflectors was initiated which eventually left the refractor far behind with respect to the size of the clear aperture.

In 1835 J. von Leibig determined how to coat glass surfaces with silver. This was a most important innovation. Later K. A. Steinheil and J. B. L. Foucault, in particular, further improved this procedure and in 1857 produced the first glass telescope mirrors with silvered surfaces. The relatively rapid adoption of silvered glass mirrors by astronomical observers is due to a number of advantages which they manifest with respect to metal mirrors and lenses. In comparison to metal mirrors, silvered glass mirrors have considerably higher reflectivity, while at the same time they weigh considerably less, and also allow a better polish. The optical quality of the glass does not have to be as high as that for lenses. Lastly, the physicist Foucault developed a highly empirical, but at the same time very simple, method for the testing of parabolic mirrors, with which deviations from the ideal surface on the order of 1/100 the wavelength of light could be measured. Foucault himself completed a mirror of diameter 80 cm according to his new method, and the 1.2-m telescope of the Paris Observatory (1878) also resulted from his initiative.

Fig. 57. Carl Zeiss

Fig. 58. The 120-cm reflector for the Babelsberg Observatory during its assembly at the Zeiss firm in Jena

Fig. 59. Large refractor of the Astrophysical Observatory of Potsdam

The scientific requirements exhibited a strong correlation with the further development of the reflecting telescope. The whole unfolding of astrophysics, spectroscopy, and photography, and the active interest concerning faint nebulous objects demanded substantial increases in light-gathering power. Thanks to the work of H. Draper, A. A. Common, and G. W. Ritchey in the first decades of the twentieth century, important progress was made in the production and mounting of larger reflectors. In 1908 a reflector with a mirror diameter of 1.5 m went into operation at Mt Wilson Observatory. Hale, who had already acquired the necessary sum for the construction of the largest refractor from the Chicago railroad magnate Yerkes, then approached a businessman from Los Angeles, J. D. Hooker, for the financing of a giant reflecting telescope with a mirror of diameter 2.5 m. The mirror's disk was ordered from France. The difficulties in the casting were considerable for such a large disk and were not successfully dealt with until the fourth attempt. After a five-year term of work the figure of the mirror was shown to be accurate enough to allow the final construction of the telescope to begin. In 1917 the giant instrument, whose movable parts had a mass of 90 000 kg, was installed on Mt Wilson for systematic observations (Fig. 60). With this the world's largest reflector went into operation. The great importance of this instrument was proven to the utmost degree in many research projects over the following decades. The large reflecting telescopes had finally come of age. The American 5.08-m reflector (operational in 1947) and the Soviet 6.00-m reflector (operational in 1974) fully confirmed this tendency, and the presently projected large instruments, about fifteen in number and international in scope, are all reflectors.

We should also mention that the progress in technology led to a number of highly specialized auxiliary instruments and tools as a direct result of various astronomical challenges. The development of these tools brought about improvements in other, non-astronomical, instrumentation. The variety of the problems, especially after the growth of astrophysics, led to the construction of spectrographs, astrocameras, and special photographic refractors and astrographs, photometers, plate-measuring machines, and other data reduction devices. However, the principal instruments themselves were modified and refined by genial designers, practitioners, and theoreticians for a wide variety of purposes. Meridian circles are noteworthy in this regard; they are crucial for the measurement of position of celestial bodies. For these the stability of the mounting is particularly important.

It was shown that the resolving power of a telescope, which is determined by the diameter of the objective, could be improved with the use of such devices as the stellar interferometer of A. A. Michelson (1920). With such a device it was possible to measure directly the separation of very close double stars, the diameter of Jovian moons, and eventually even the diameters of several red giant stars.

Fig. 60. The Hooker reflector (100-inch) of Mt Wilson Observatory

Fig. 61. Dome of the large refractor of the Astrophysical Observatory of Potsdam

Fig. 62. Micrometer of the large refractor of Pulkovo Observatory

Differing instrument types were developed purely by tradition for the investigation of the Sun. The great luminosity of the Sun allows the use of a long focal length. A first instrument of this type was the Snow telescope of Mt Wilson Observatory (1904/05). The coelostat, as astronomers have used it since the beginning of the nineteenth century, projects the light of the Sun onto a horizontal reflecting telescope. As a result, however, the important condition of constant temperature of the individual optical components cannot be achieved. Subsequently, due to the substantial effort of Hale, the tower telescope for solar research was developed in the years 1906–1908. The coelostat was then placed on a high tower where it directs the light into a vertical telescope, whereby the spectrograph is kept in a vibration-free room at constant temperature. The Einstein Tower of the Astrophysical Observatory of Potsdam, designed and built between 1920 and 1924, was laid out in a similar way, but with a horizontal spectrograph room. It was also known for its architecture.

Theoretical and practical opticians devoted particular attention to the further improvement of the reflecting telescope and to the ever better adaptation to the respective scientific objectives. Therefore, among other things it was necessary to try to eliminate coma from the imaging of celestial bodies (which results in optical systems with parabolic mirrors). The rays do not travel parallel, but are inclined toward the optical axis of the parabolic mirror; instead of a point of light at the focus, a smeared-out

Fig. 63. Plate-measuring machine of Gautier for the determination of stellar coordinates

image appears. This phenomenon is called coma and it increases in proportion to the square of the clear aperture of the mirror. This becomes quite a considerable inconvenience for large-diameter telescopes. In particular, K. Schwarzschild tried to get around this weak point of design by means of ingenious considerations. He developed a mathematical theory of the reflecting telescope.

However, B. Schmidt, the genial optician and amateur astronomer from Mittweida, made the most efficacious contribution to the development of a coma-free mirror system.

Fig. 64. Spectroscope of the Toepfer firm, with which H. C. Vogel first determined the radial velocities of stars

In 1931 Schmidt employed a spherical mirror for the construction of a coma-free mirror system; at the center of curvature of the spherical mirror Schmidt placed a diaphragm opening. The image is coma-free in any case, independent of the direction from which the rays pass through the diaphragm toward the mirror. The image surface upon which the images result is spherical. The center of curvature of this image surface is identical to that of the primary mirror. For the elimination of spherical aberration Schmidt then introduced a correction plate having a complex figure, which is placed at the position of the diaphragm opening. The Schmidt telescope allowed large fields of view to be photographed without image defects.

Fig. 65. Meridian circle of Reichenbach (1819)

The amazing simplicity of the solution to the problem has led in the meantime to a worldwide distribution of the Schmidt system. Later, other attempts were made to obtain similar success with comparable means. Above all, the solution of the Soviet optician D. D. Maksutov is well known. In a Maksutov reflector (1944), instead of a corrector plate, a meniscus lens is used for the elimination of spherical aberration.

Fig. 66. Original spectroscope of the Steinheil firm of Munich, with which G. Kirchhoff carried out his investigations of the solar spectrum

Table 8. *Some large refractors of the nineteenth and twentieth centuries*

Location	Year of installation	Manufacturer	Objective diameter and focal length in centimeters
Dorpat	1824	Fraunhofer	24.4/413
Pulkovo	1839/40	Merz & Mahler	38/690
Cambridge, Mass.	1843	Merz	38/690
Washington, DC	1871	Clark	66/990
Vienna	1878/79	Grubb	68/1054
Pulkovo	1885	Clark/Repsold	76/1406
Mt Hamilton (Lick Observatory)	1888	Clark	91.4/1828
Flagstaff, Ariz.	1896	Clark	61/945
Berlin-Treptow	1896	Steinheil/Hoppe	68/2100
Williams Bay, Wisc. (Yerkes Observatory)	1897	Clark/Warner & Swasey	102/1979
Potsdam	1899/1905	Steinheil/Repsold	80/1200 50/1250
Berlin-Babelsberg	1913	Zeiss	65/1050
Tokyo	1930	Zeiss	65/1050 35/1050

OBSERVATORIES

The enormous explosion of astronomical research in the nineteenth century is inseparably connected with the establishment of the laboratories of astronomy – the observatories. Never before were so many observatories created in such a short span of time as during the nineteenth century.

The observatories are the result of, as well as the reason for, the development of astronomy. The establishment of new observatories depends on existing economic and scientific requirements and possibilities. For example, many observatories owe their origin to astronomical geography, although their spheres of activity later assumed a considerably wider development. In Russia, for example, the growth of an astronomical research tradition did not begin until the time of Peter the Great; before that Russia did not control any sea-shores. After this the development of Russian astronomy was substantially inspired by navigational schools.

Without the scientific frontier of astrophysics many well-known observatories would likewise not have arisen. But a subsequent development of astronomy would not have been possible without these new establishments.

Before the nineteenth century only about three dozen important observatories existed. A century later there were more than 200, along with a large number of smaller stations, some of which only existed for a short time. In the nineteenth century the founding of observatories was the result of a classical organic evolutionary process. A statistical investigation concerning the foundings of observatories by this author showed that the number of observatories worldwide increased exponentially with time (Fig. 67),[1] as is

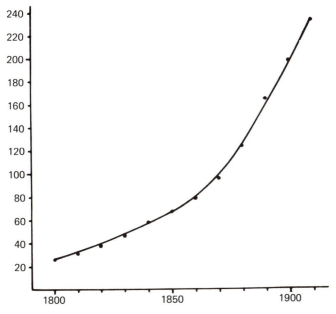

Fig. 67. Increase of the number of observatories for the period of time 1800–1910

Fig. 68. Babelsberg Observatory (today run by the Central Institute for Astrophysics of the Academy of Sciences of the GDR)

Fig. 69. Kazan University Observatory in 1842

Fig. 70. Dorpat Observatory in the nineteenth century

Fig. 71. Einstein Tower for Solar Research in Potsdam (Academy of Sciences of the GDR)

the case for other growth processes like those in physics and biology. This is undoubtedly the result of a fundamental underlying mechanism, rather than a conscious effort on the part of those who established new observatories. It is amazing how sensitively various countries reacted to the growing needs of astronomy and to what degree corresponding measures were taken.

The founding of observatories took place very differently in the individual countries. Understandably, the richer countries could afford better-equipped observatories. During the seventeenth and eighteenth centuries France and England had those observatories which were unsurpassed in technical capability; during the nineteenth century Germany's observatories surpassed those of France and England. In the second half of the century, however, the USA moved into the lead. With the emergence of astrophysics, astronomy, more than ever before, represented a fundamental science, the immediate uses of which were not fully known; important scholars often had to bring the whole weight of their authority to their governments and now and then also had to introduce overstated practical arguments, in order to obtain the funding for the large institutions. An example of this is W. Foerster's detailed memo to the German crown princes, in which he described more than anything the importance of the Sun for the activities on the Earth, and in which he described the large-scale scientific research of the Sun as a central concern of high practical importance. In this way the ground was laid for the establishment of the Astrophysical Observatory of Potsdam.

In the United States of America it was the pursuit of prestige that led to the origin of new observatories. In 1825 President John Quincy Adams referred to the great number of European observatories, and how the USA had nothing by comparison. The extremely rapid changes in America from the Revolutionary War to the middle of the nineteenth century, which exhibited more than anything greatly increased industrial production and which was accompanied by a notable increase of population (8.4 million people in 1815; 31.5 million in 1860), created just the right economic conditions for observatories to spring up all over the USA. The first state astronomical undertaking, the US Coast Survey, did not arise by chance. Indeed, the instruments at that time came from the most famous European instrument makers, and the director, F. R. Hassler, was Swiss. Not until later did the USA have its own instrument makers and suitable technical capabilities. Russian astronomy in the nineteenth century had the control of relatively few, but on the other hand, very well-equipped observatories, in particular the observatories at Dorpat and Pulkovo. These world famous research institutions achieved significant contributions to the development of astronomy in Russia. However, this science did not take a large step forward until after the Great Socialist October Revolution. Immediately after the Revolution, because of the difficult economic and political problems the Soviet Union had to endure, Lenin emphatically stressed the challenge of science, and even astronomy by name. Today in the Soviet Union there exists a total of 47 observatories; of the large observatories of international stature operating today in the USSR, about half were founded after the October Revolution. The old observatories were in part brought up to date with new instruments and considerably enlarged in their research capabilities.

While the observatories in the nineteenth century had a corner on the market of astronomical knowledge, things are different in the twentieth century. There has been a continuous increase in the number of contributions from theoretical institutions, related sciences, and eventually from interplanetary flights. Today the amount of astronomical research can no longer be estimated exclusively by the number of existing observatories.

CONGRESSES AND ASSOCIATIONS

With the growing number of observatories, astronomers, and areas of research, the problem of close communication among researchers for the mutual planning of programs of work became more and more central. Farsighted scholars became familiar with this necessity and tried to accomplish a creative contribution to the development of astronomy through proper measures.

F. X. von Zach was an outstanding personality and a genuine motivator of the development of international astronomy at the beginning of the nineteenth century. He was highly respected in the international scholarly world. He had the ability to sense how research was progressing and how it should progress, and had either met all important astronomers or corresponded with them. Zach represented a new type of scholar; he was a science organizer. One of his aphorisms was: 'Great things can only be achieved through the union of many talents.'[2] He sought to bring about this union on an international scale. It was clear as a result that he had to accomplish difficult pioneering work. Many national, scientific, and personal prejudices of the scholars had to be resolved. The awkward situation in Germany (which was not a unified country at that time) came into play; this was a significant obstacle which interfered with the task at hand. Zach suggested that because Germany did not comprise a nation, there lacked centrally directed scientific institutions on par with the English Board of Longitude. Time and again the publication of important books and journals came to nothing because the publishers were concerned only with their own business interests. Because of this adversity Zach, with extraordinary energy, set into motion a series of projects which in total played a notable role in the rapid growth of astronomy in the first decades of the nineteenth century. He founded the first astronomical journals, organized scientific congresses, founded a scientific society, and devoted the greatest attention to the rising generation of scientists.

Later, due to the social conditions of fully developed capitalist relationships, and at a progressive scientific level, other outstanding personalities appeared who built upon these beginnings and who were even able to get successful international investigations off the ground. In particular, we should mention the American astronomer and organizer, G. E. Hale, and the German astronomer and organizer, W. Foerster.

The particular aptitude of both men for organizational work rested on their thorough technical backgrounds, similar to Zach and others. Hale in particular acquired an international reputation as a pioneer of solar research. Concerning his scientific work he was a consistently energetic organizer. Given the particular situation in America, he understood how to approach institutions and wealthy individuals for the financing of expensive projects like the construction of Yerkes Observatory and the Hooker reflector at Mt Wilson Observatory. The 5-m reflector at Mt Palomar (the Hale telescope) was also the result of his initiative, even though the installation of the instrument did not take place until after his death. Hale founded the *Astrophysical Journal* and introduced international collaborative work in the area of solar research through the creation of the International Union of Solar Research. As Foreign Secretary of the National Academy of Sciences (NAS) he also played an important role in the development of astronomy throughout the USA.

Foerster supported the founding of the Astronomische Gesellschaft and the organization of the International Earth Survey, directed the adoption of the metric system in Germany, and did a great deal to facilitate the founding of the Astrophysical Observatory in Potsdam; he busily involved himself with the establishment of the Jena glassworks and the construction of the National Institute of Physical Technology. The founding of the public time service and the society Urania in Berlin were also the result of his initiative. Finally, it was also Foerster who proposed the Central Bureau for Astronomical Telegrams and brought to life the International Latitude Service. This list of accomplishments could only have been carried out with great know-how, talent, and a view for the whole development of astronomy.

The unfolding of astronomy was no longer possible without the purposeful organization of research. Congresses and associations, as well as the creation of specific technical periodicals, are therefore an inseparable component of the history of astronomy of that time period.

The collaboration of individuals made possible work of a new quality, in which the contributions of individuals did not simply sum together; rather, their efforts inspired a whole range of projects, the scope of which was previously unimaginable.

Though research is individually planned and carried out, the face to face meeting of scientists plays a significant role in the close interplay of different kinds of research. Zach, who had met many astronomers around the world, knew the merits of such contacts and made a concerted effort in this regard;

Fig. 72. F. X. von Zach

he invited numerous astronomers to a meeting at the Duchal Observatory in Gotha during the summer of 1798. According to its format, content, and purpose, the meeting of fifteen European scientists which came about as a result can be regarded as the first organized astronomical congress of the more recent history of astronomy. Because of the fruitfulness of the day-long technical discussions, which even precipitated a scientific resolution, many participants hoped that a similar meeting could take place as soon as possible. On the other hand, one must also understand that the prevailing political conditions in Germany did not provide a favorable climate for such projects; simply the participation of the important French astronomer Lalande brought about intense feelings of mistrust on the part of some of the astronomers, but even more so in the 'trembling and shaking little German princes'.[3] A number of the invited astronomers stayed away from the meeting for this reason, and the Duke of Gotha even tried to enact a sovereign restraining order against the meeting with respect to the English government. The fear of the spread of revolutionary thinking was great. Schroeter spoke of a 'French astronomical regime'[4] and complained with other astronomers about the metric system, a 'child of the French Revolution', which according to his opinion could only add fuel to the fire of French imperialism. It was even more difficult at that time to create an energetic astronomical association in disunited Germany. In 1800 Zach and Schroeter attempted to found a United Astronomical Society, with whose help actual scientific research could be planned among the astronomers. However, nothing came of the fruitful activity of this society because of the political situation in Europe at that time.

In contrast to this in England the Royal Astronomical Society arose in 1820. It established its own public organ in 1831 – the *Monthly Notices*. This society proved to be a noticeable inspiration for the development of astronomy in England.

Of the German organizations only the Gesellschaft Deutscher Naturforscher und Ärzte (Society of German Naturalists and Doctors, founded in Leipzig in 1822) enjoyed stability. Its founder, L. Oken, demonstrated that the German naturalists had to organize themselves if, for example, German science wanted to keep pace with England and France. On the occasion of the assembly of the Society in 1828 in Berlin a section for geography and astronomy was also established. Thirty years later this section decided to create a separate astronomical union. A total of twenty-five astronomers got together in Heidelberg in 1863 for the founding of the Astronomische Gesellschaft. With this a technical union arose, whose initiators attempted from the very start to enlist foreign astronomers for the activity of the society. Eventually nearly all active astronomers of the world were members; the membership rose from 173 in 1865 to more than 500 in 1930. Thus several undertakings carried out worldwide were accomplished, of which only the zonal catalogs of the AG have been mentioned (see p. 32).

Other countries organized their own astronomical organizations. In Italy there was the Società degli spettroscopisti Italiani, in Russia the Russische Astronomische Gesellschaft in St Petersburg (1890), in France the Societé astronomique de France (1887), in Canada the Royal Astronomical Society of Canada (1890), and in the USA the Astronomical and Astrophysical Society of America (1899). All told during the nineteenth century about twenty national astronomical organizations were founded, whose collective goal was an intensification of research. In the first decades of the twentieth century this tendency continued.

However, the scope of problems grew considerably, and the amount of specialization soon required a more effective international collaborative effort on the part of astronomers. At the beginning of the nineteenth century hardly any of the astronomers were specialized. Most of them concerned themselves at the same time with several special areas. At the beginning of the twentieth century there were thirty times as many astronomers (approximately 2000 astronomers in 1900). The majority were devoted to very specialized research. If we think a little about the beginnings of astrophysics, then there are only a few specialists for given questions within the bounds of a national astronomical association. As a result the problem of collaborative work with the specialists from different countries moved into the foreground. The most consequential expression of this necessity was the founding of the International Astronomical Union (IAU) in 1919. The national astronomical organizations are fused together in it, while the different specific problems are worked on in a great number of technical commissions.

The international collaborative work did not take shape without conflict; it is particularly dependent on social development. The strong concentration of international scientific societies in Germany at the beginning of the twentieth century makes this very clear. It was due to a great degree to the achievements and the initiative of progressive German scientists. However, the politics of German imperialism and militarism brought this work to nothing within a short time. The defeat of German imperialism in the First World War led to a total boycott of German science; this even found its place in the Versailles Treaty. All central bureaus which had been set up in Germany were re-established; German scholars were struck from the rolls of international organizations; areas of work for which the German collaboration was unconditionally required were dissolved; works of German scientists were not considered in international bibliographies; Germany was excluded from international congresses, etc. It is only fair to mention that the 'Summons to the Cultural World' of October 1914, signed by 93 world-renowned German intellectuals, had a lot to do with this boycott; it called the advance of German imperialism good and even pictured Wilhelm II as the 'protector of freedom'. The damage which this manifesto caused to the development of science could not be undone by a little-known counterproposal composed by Einstein and the Berlin physiologist G. F. Nicolai, and countersigned by W. Foerster.

This example, which is just one of many like it, shows that the struggle for scientific success cannot be detached from social progress. Particularly after the bitter experiences in the twentieth century this is known to an ever increasing number of scientists. They do not fear to leave temporarily the realms of 'pure science' in order to express their scientific principles against reaction and for social progress.

ASTRONOMICAL LITERATURE

Literature is an unbiased reflection of the development of a science. It is the result of, and at the same time the reason for, scientific work.

As long as the number of active astronomers was small the scope of the resultant knowledge also remained small. The results were set down in books which in general represented sufficiently standard works for quite a long time. Besides, scholars had access to the publication organs of the academies. The French *Journal des Savants*, the English *Philosophical Transactions*, and the German *Acta Eruditorum* are the oldest and best known non-specialized academic journals. The academies – themselves products of bourgeois interests – made these publication organs more and more the platform for the sciences with practical applications.

The scholars exchanged the rest of their results and opinions among themselves by means of correspondence. The participation in a 'correspondence circle' was an important means for young scholars in the seventeenth and eighteenth centuries to become known in scientific circles.

At the beginning of the eighteenth century the number of astronomers was so small that this form of communication was effective. The tremendous expansion of research, however, at the turn of the nineteenth century caused a continuous increase in the number of astronomers. With this the time was ripe for the birth of astronomical technical journals; for now it was impossible for individual astronomers to regularly maintain the required correspondence with all active astronomers. The expenditure of time necessary for this diminished the effectiveness of the immediate research. In 1774 the *Astronomisches Jahrbuch* (Astronomical Yearbook) started appearing, having been founded by J. H. Lambert. From 1777 to 1829 it was edited by J. E. Bode. From its first appearance it primarily published ephemerides, but there were a number of treatises which contained results of international astronomical research. The yearly issuance of the *Jahrbuch* was clearly not suitable for having information available on a short time scale. Therefore, the farsighted F. X. von Zach created a specialized technical journal in 1798, the *Allgemeine Geographische Ephemeriden* (General Geographical Ephemerides). The active promotion of research was the main purpose of this journal. Zach therefore favored the original scientific treatise and promoted the debate of opinions. These progressive reasons – at that time not at all common for journal editors – contributed to rapid progress in research. Barely two years later more determinations of position were being published than ever before, and work continued for newer, more accurate measurement procedures and calculating procedures.

Indeed, the *Allgemeine Geographische Ephemeriden* were a bit specialized. In addition to geographical contributions, which prevailed percentage-wise, there were also many ethnological and meteorological contributions, such that only about 25 per cent of the contributions were astronomical in nature.

A direct continuation of the *Ephemeriden* was the journal *Monatliche Correspondenz zur Beförderung der Erd- und Himmels-Kunde* (Monthly Correspondence for the Promotion of Geography and Astronomy), which first appeared in 1800; it was edited first by Zach, then by B. A. von Lindenau, who was later the Minister of State of Saxony. Particularly under Lindenau's auspices the journal became one of the most important public organs for astronomical research in the first quarter of the nineteenth century. To a great degree the journal met the demands of scientific research, as shown by the extremely short time span between date of submission and the appearance of important contributions (Fig. 73). Unfortunately, this meritorious journal met a quick end in the course of the War of Liberation after only fourteen years of existence. Lindenau, a spirited patriot, took part in the fight against Napoleon's troops, despite the fact that as an astronomer other suitable activities would have existed for him.

When Lindenau returned to Gotha in 1814 he had a plan for a new journal. It first appeared in 1816 and was coedited by him and Bohnenberger. This was the *Zeitschrift für Astronomie und verwandte Wissenschaften* (Journal for Astronomy and Related Sciences). This journal was a great step ahead toward a specifically astronomical technical journal. On average 70 per cent of the contributions dealt with astronomy. However, Lindenau's unavoidable transfer into state service, which at the same time brought about his final departure from astronomy, led to the discontinuation of the journal after three years. The journals at that time were similar to the private observatories in that their effectiveness strongly depended on the personal initiative of individual scholars. The low degree of institutionalization of these projects was the reason for the lack of continuity. In Germany it was particularly evident that there was a great need for a central organization which was strong politically and scientifically, like those which existed in England and France. There, science was used purposely for the good of the state; in the German states outstanding pioneers often came up empty-handed due to the prevailing conditions. The difficulties which were encountered with the publication of such important works as Bessel's *Fundamenta Astronomiae* and the *Theoria Motus* of Gauss are prime examples of how to actually hinder the advance of science. It was only due

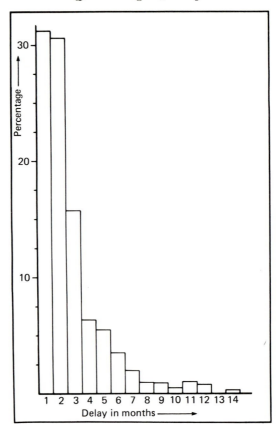

Fig. 73. Publication delay times for 521 dated contributions in the journal *Monatliche Correspondenz zur Beförderung der Erd- und Himmels-Kunde* (1800–1813)

to the exemplary intervention of Zach and Lindenau and the support of Duke Ernst II of Saxe-Gotha that these two works could be published. The Berlin Academy itself withdrew support of the *Astronomisches Jahrbuch* for financial reasons, and Bode was forced to continue this work, known worldwide, as a private undertaking.

In 1821 a real astronomical journal was finally created – the *Astronomische Nachrichten* (*AN*, Astronomical Notes; Fig. 74).* It was edited by H. C. Schumacher in Altona and was supported financially by the Danish court. (The Danish Finance Minister von Mösting, a well-known amateur astronomer and promoter of science, was particularly enthusiastic about

* The first issue of the *Astronomische Nachrichten* was published in 1821. The first volume was completed in 1823, as a result of which there has been some confusion as to the official founding date. It is 1821. See Dieter B. Herrmann, *Die Entstehung der Astronomischen Fachzeitschriften in Deutschland (1798–1821)*, *Veröffentlichungen der Archenhold-Sternwarte*, Berlin-Treptow, no. 5, 1972, p. 84 and p. 128. (Tr. note)

ASTRONOMISCHE

NACHRICHTEN

herausgegeben

von

H. C. Schumacher, Ritter vom Dannebrog,

ordentl. Professor der Astronomie in Copenhagen, Mitglied der Königlichen Gesellschaften der Wissenschaften in Copenhagen, Neapel, London und Edinburgh, der astronomischen Gesellschaft in London, und der Königlichen Landhaushaltungsgesellschaft in Copenhagen, Ehrenmitglied der Society of useful arts in Edinburgh.

Erster Band.

mit 3 Kupfern, 9 Beilagen, und einem Register.

Altona 1823.

gedruckt in der *Hammerich*- und *Heineking*'schen Buchdruckerei.

Fig. 74. Title page of the journal *Astronomische Nachrichten*, vol. 1, Altona 1823

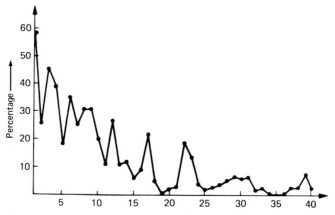

Fig. 75. Distribution of the proportion of contributions concerning astronomical geography in the journal *Astronomische Nachrichten*, vols. 1–40

the new journal.) The *Astronomische Nachrichten* quickly moved to literary center stage for astronomers throughout the world. The individual numbers of the journal appeared without constraint, always according to the extent of the material in question, such that the information arising in research could be immediately distributed. Right from the start the contents became more and more astronomy oriented (Fig. 75). (Some of the work was more along the lines of technical journalism.) Thus over many decades the *Astronomische Nachrichten* had no serious competition. Essentially all important astronomical discoveries after 1821 were described in the pages of this journal. Only in England was there a comparable astronomical publication – the *Monthly Notices of the Royal Astronomical Society*; this journal, however, did not have the international flavor of the *Astronomische Nachrichten* because of its connections with the English astronomical community. Indeed, the rapid development of astronomy in many countries later led to the origin of other technical journals. For example, the *Astronomische Nachrichten* was a prototype for the *Astronomical Journal* in the USA (1849).

To be sure, astronomical knowledge was not only published in journals. Moreover, individual observatories also published their own contributions. In this way about 200 observatory publication series arose between 1850 and 1900.

The monograph generally lost its importance as the form of primary scientific literature; instead, original publications appeared in journals. Monographs contained summaries of whole sciences or individual disciplines. Typical examples of this format are Secchi's book *Le Soleil* (The Sun) and Müller's *Photometrie der Gestirne* (Photometry of the Stars).

The *Astronomischer Jahresbericht* was founded in 1899 by W. F. Wislicenus in order to facilitate access to sought-after information in the great variety

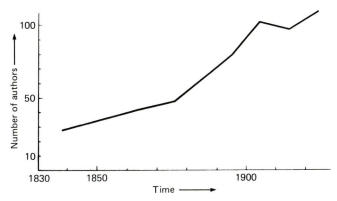

Fig. 76. Distribution of the number of authors per year in the journal *Astronomische Nachrichten*

of scientific periodicals. For the time period of report of one year it contains a review of the scientific literature of astronomy and its related subject areas. Notable scientists throughout the world worked on the *Astronomischer Jahresbericht*. As a consequence of the rapid growth of astronomy this reference source was taken over in 1969 by the *Astronomy and Astrophysics Abstracts*, which appears semi-annually.

In the nineteenth century popular scientific literature also developed in the area of astronomy. Today it is widely read throughout the world, for astronomy has made great contributions to the world's ideological views.

During the European Enlightenment the popularization of science was already shown to be important for the growth of knowledge. Such prominent scholars as Francois Arago and Alexander von Humboldt advocated broad dissemination of astronomical knowledge. The popular books by Littrow, Bode, Mädler, and Diesterweg – some of which were published long after the authors had died – taught the public a great deal about astronomy. S. Newcomb's *Popular Astronomy* was one of the most popular astronomical books. Since 1910 it has appeared in seven languages. The German translation, for example, went through many editions. On the basis of this book Newcomb composed other popular scientific works like *Astronomy for Everybody* and *Sidelights on Astronomy*. C. Flammarion's *Astronomie populaire* was also widely distributed, as was the book by F. Arago with the same title, which also appeared in German translation. Many other scientists, among them astronomers of world renown, wrote popular astronomical books.

Astronomy is unique among the natural sciences in that it has its own genuine institutes of popularization (the public observatories), many astronomy clubs, and truly popular journal literature. G. A. Jahn published the *Unterhaltungen im Gebiete der Astronomie, Geographie, und Meteorologie* (Investigations in the Areas of Astronomy, Geography, and Meteorology); it appeared weekly starting in 1847 and ran for ten years. In 1860 C. A. F.

Peters, together with K. Pape, began publishing the *Zeitschrift für populäre Mitteilungen auf dem Gebiete der Astronomie und verwandter Wissenschaften* (Journal for Popular Contributions in the Area of Astronomy and Related Sciences), of which, however, only a few volumes appeared. A successful popular astronomy journal was *Sirius* (since 1868), which in 1927 merged with the journal *Die Sterne* (The Stars, founded in 1921). It is still published today in Leipzig. Three other German popular scientific magazines also achieved wide distribution: *Himmel und Erde*, which appeared from 1889 to 1915 as the journal of the society Urania; *Die Himmelswelt* (1920–1950); and the monthly journal *Das Weltall* (1900–1944), which was published by the Treptow Observatory in Berlin.*

Today astronomy has a widely diversified and well-built net of publications. The tendency of previously national journals to merge is realized by international publication organs in order to facilitate access to the literature. Each year approximately 10 000 titles with astronomical contents are published. However, knowledge is still on the upswing; the increase, as measured by the number of papers published, amounts to about 4 per cent per year. As a result astronomical publications face challenges similar to those found in all realms of scientific knowledge. We are finding new ways to store and retrieve information. Data bases which are accessed by computer are playing an ever more important role. This may make journals less important as time goes on.

* A much-read popular journal in English is *Sky and Telescope*, which first appeared in 1941 as a merger of *The Telescope* (founded 1933) and *The Sky* (founded 1936). It is published in Cambridge, Massachusetts. This periodical is read by many professionals as well. (Tr. note)

Afterword

The hundred and fifty years of intensive research from the late eighteenth century to the early twentieth century fundamentally changed our conception of the universe. In Hertzsprung's day research had more than achieved what the most daring fantasies in Herschel's day had ever predicted.

A whole range of new phenomena had been discovered. Various effects which had previously been suspected could be quantitatively confirmed through the application of fundamental laws. While positional astronomy in Herschel's day only penetrated about 160 light-minutes into the cosmos, the contemporaries of Hertzsprung had measured some objects at distances of millions of light-years. While at the beginning of the nineteenth century the validity of the laws known at that time was in doubt for distances on the order of the size of the solar system, in the twentieth century it was shown with certainty that the laws apply throughout space where the same conditions are met. The material unity of the universe, as the astrophysical research procedures had shown, was one of the most sublime results of astrophysics and a great enhancement of the Copernican paradigm of the unity of the Earth and celestial bodies.

Two hundred years ago daring philosophers and naturalists postulated, but could not prove, that the universe was evolving. Now this is an undeniable fact. The problems of the birth and death of planets, suns, and stellar systems became a central concern of research.

Often astronomical research in the epoch from Herschel to Hertzsprung was pushed to seemingly insurmountable limits; however, no individual theory is really invincible. The practice of research has refuted all assertions concerning the existence of assumed barriers to knowledge.

With every problem solved, however, new questions arise. Every discovery is only one more contribution in the iterative and asymptotic process of finding scientific truth. In science a situation is never reached, 'where one cannot go further, where nothing more is left to do than to place your hands in your lap and gaze at the absolute truth achieved'.[1] As a result new starting points often result from surprising developments, which in part concern other sciences. The application of astrophysics is an impressive proof of this, not only with its results, but rather also with its technical procedures. The development of astronomical research in the last few decades has shown this even more clearly. Nearly all fundamentally new results are based on the further application of numerous achievements of science and technology in their widest scope.

The question of the energy sources in the stars had always been a difficult problem. Helmholtz's hypothesis that stellar energy is produced by the contraction of the stars was generally accepted. After the discovery of radioactivity, with which an absolute age scale of the Earth's history could be made, it was shown, however, that this hypothesis made the stars much too short lived. Nuclear physics finally provided the solution. In the years 1937/38 H. Bethe and C. F. von Weizsäcker were able to prove that nuclear fission takes place in the solar core – protons are transformed into helium nuclei after several intermediate steps (the Bethe–Weizsäcker, or proton–proton, cycle). This told us more than just how to calculate the ages of the stars; it primarily provided the basic assumption for theories of stellar evolution. Now it was possible to describe mathematically the physical events and conditions and to represent evolution theoretically by means of the calculation of stellar models for different points in time. Computers play a great role in these very extensive and complicated studies. Previously, the physical conditions had to be considered in simplified form in order to derive numerical answers in a reasonable amount of time. As it turned out the energy production of stars is associated with a comparatively very small mass loss. During the entire long phase of life, which is called the hydrogen burning phase, the stars maintain their positions on the main sequence of the Hertzsprung–Russell Diagram. After the exhaustion of hydrogen fuel the conditions of state change in a characteristic way. The stars eventually move away from the main sequence. The old simplified interpretation of the H–R Diagram as an evolutionary diagram had thus been shown to be inapplicable. Moreover, it was shown that it is really a diagram of state in which the number of stars in each part of the diagram directly indicates the relative time scales for different stellar evolutionary processes. Indeed, over the course of their lifetimes the stars trace out specific paths in the H–R Diagram, but not in the simple, previously assumed way.

Completely new knowledge resulted from the reception of non-optical electromagnetic waves from the cosmos. This was made possible by technical development.

At the beginning of the 1930s Jansky discovered the cosmic radio waves when he determined that the noise reception of his very high frequency wave receiver varied with a period of one sidereal day. A new branch of astronomy – radio astronomy – arose after the Second World War as a result of the great progress that had been made in radio technology. Today radio astronomy represents the best developed part of non-optical astronomy and it has given us important results concerning the study of the cosmos.

Radio astronomy also has been able to provide greater understanding of the large-scale structure of the Milky Way system. The spiral structure of our galaxy can be mapped out by observing at the 21-cm line of neutral hydrogen. Many results widely discussed today are likewise due to this branch of modern astronomy: the discovery of quasars (1960), pulsars (1967), the discovery of the 3 K background radiation (1965), and the detection of many complicated organic molecules in interstellar space.

The 'radio window' is not the only non-optical window to the universe. In particular, the results of space travel after the opening of the cosmic age by the Soviet Union in 1957 have created new possibilities for astronomy. In recent times X-ray astronomy, infrared astronomy, and extraterrestrial astronomy have undergone rapid development. Hand in hand with powerful new telescopes on the Earth, which are presently in operation or will be erected in the near future, astronomers systematically proceed with the inquiry into the complicated processes in the cosmos. A finely articulated, widely branched, interdisciplinary line of research has developed from the once monolithic block which astronomy represented at the time of Herschel. A modern bibliography would list more than 150 special areas of present-day astronomy research. New secrets are constantly being wrested from the cosmos in unrelenting research work.

The cosmos is for research a powerful laboratory in which the events of nature play their part. It cannot be simulated in the terrestrial laboratories to any comparable degree. The investigation of these processes will also help us in the future to know other laws of nature, to use them for our needs, and to understand our proper place in the universe.

Chronology

1781 William Herschel discovers the planet Uranus on 13 March.
William Herschel begins his systematic studies of the 'construction of the heavens', as well as his observations and cataloging of nebulae.

1782 William Herschel publishes his first double star catalog.
William Herschel publishes his discovery of the motion of the Sun.

1784 William Herschel publishes his first investigation concerning the structure of the Milky Way system.

1785 Goodricke discovers the brightness variations of δ Cephei.

1786 Pigott publishes a first catalog of variable stars containing twelve objects.

1789 William Herschel publishes his nebula classification.

1794 Chladni proves the extraterrestrial character of meteors.

1795 William Herschel publishes his views on the nature of the Sun and fixed stars.

1796 Laplace publishes his cosmogonic theory in the book *Exposition du système du monde* (Treatise on the System of the World).

1797 Publication of Olbers' 'Treatise concerning the easiest and most convenient method of determining the orbit of a comet.'

1798 Brandes and Benzenberg ascertain the distance of meteors.
Fifteen European astronomers get together for technical discussions at the Seeberg bei Gotha Observatory (Gotha Astronomical Congress).
In Gotha, Zach begins publishing the *Allgemeine Geographische Ephemeriden*.

1800 William Herschel discovers infrared rays in the solar spectrum.
Zach begins publishing the *Monatliche Correspondenz zur Beförderung der Erd- und Himmels-Kunde*.

1801 Piazzi discovers the first asteroid, Ceres, on 1 January.

1802 Olbers discovers the second asteroid, Pallas.
Wollaston uses a slit for the production of solar spectra.

1803 Lalande publishes his *Bibliographie astronomique*.

1804 Harding discovers the third asteroid, Juno.
In Munich an optical–mechanical institute is founded by Fraunhofer, Utzschneider, and Reichenbach.

1807 Olbers catalogs the fourth asteroid, Vesta.

1815 Fraunhofer catalogs the dark lines in the solar spectrum.

1816 Lindenau and Bohnenberger found the *Zeitschrift für Astronomie und verwandte Wissenschaften*.

1818 Publication of Bessel's *Fundamenta Astronomiae*.

1820 The Royal Astronomical Society is founded in London.
1821 Schumacher founds the *Astronomische Nachrichten*.
  The Catholic Church lifts the ban on teaching Copernicanism.
1824 Lohrmann publishes his great lunar map.
1827/28 In Berlin von Humboldt holds his popular lectures on the cosmos.
1828 In Berlin the Society of German Naturalists and Doctors organizes a section for geography and astronomy.
1833 In Göttingen the first magnetic observatory is founded.
1834 Weber and Fechner discover the fundamental law of psychophysics, and Fechner applies it to stellar magnitudes in astronomy.
1837 Wilhelm Struve successfully measures the parallax of the star α Lyrae.
1838 Bessel determines the trigonometric parallax of 61 Cygni.
1839 Founding of Harvard College Observatory in Cambridge, Massachusetts.
  Founding of Pulkovo Observatory in Russia.
1841 Bessel determines the dimensions of the Earth with geodetic degree measurements.
1842 Doppler describes the principle of apparent frequency change of waves as a result of the relative motion of the wave source with respect to the observer.
1843 Schwabe discovers the periodicity in the frequency of appearance of sunspots.
1845 Lord Rosse puts his great reflector into operation.
1846 Galle discovers the planet Neptune on the basis of the perturbation analysis of Leverrier. The theoretical work was also done independently by the Englishman J. C. Adams.
1849 Gould publishes the first American astronomical periodical, the *Astronomical Journal*.
1851 Lamont discovers the periodic variations of the Earth's magnetism.
1852 R. Wolf determines the sunspot period to be 11 years.
1854 Arago publishes his *Astronomie populaire*.
1856/57 Liebig, Steinheil, and Foucault produce the first silver-coated telescope mirrors.
1857 Hansen publishes his lunar tables.
1860 Kirchhoff and Bunsen found spectral analysis.
1861 Zöllner establishes the foundations of modern visual photometry with his astronomical photometer.
1863 The Astronomische Gesellschaft is founded in Heidelberg.
1864 Huggins discovers emission lines in various spectra of nebulae.
1866 Secchi introduces his spectral classification for stars.
1868 Discovery of helium in the spectrum of the solar chromosphere.
  Lockyer and Janssen develop the prominence spectroscope.
1869 Beginning of the International Earth Measurement.
1870 Lane publishes his work on the theory of the Sun's temperature.
1871 Founding of the Società degli spettroscopisti Italiana.
1874 The Astrophysical Observatory of Potsdam begins its work.
1877 Schiaparelli discovers the 'canals' of Mars.
1878 Newcomb publishes improved lunar tables.
  J. F. J. Schmidt publishes the last great lunar map on the basis of visual observations.

1887 International Congress for Astrophotography in Paris.

1887–1892 Houzeau and Lancaster publish the three volume *Bibliographie générale de l'astronomie*.

1888 Küstner discovers the oscillations of the pole.

1889 Pickering discovers the first spectroscopic double star ξ Ursae Majoris.

1890 Pickering and Miss Fleming classify stellar spectra on an alphabetic system. Founding of the Russische Astronomische Gesellschaft.

1892 Vogel measures the first spectroscopic radial velocities of stars.

1893 Hale develops the spectroheliograph.

1895–1897 Rowland publishes his photographic atlas of the solar spectrum.

1896 Photographic lunar maps are first published in Paris.

1897 Miss Maury refines the spectral classification on the basis of the sharpness of lines.

1898 Newcomb publishes a more accurate value for precession.

1899 Miss Fleming determines that the star RR Lyrae is variable.
Establishment of the International Latitude Service.
Establishment of the *Astronomischer Jahresbericht* (bibliography of world literature).

1900 Keeler photographs numerous nebulae and finds spiral structure in some.
Planck determines the radiation law of black bodies and founds quantum physics.

1901 Kapteyn determines the distribution of stars by means of statistical methods.
Miss Cannon introduces spectral subclasses for stars and in essence completes the Harvard classification scheme.

1904 Hartmann discovers the 'stationary calcium lines'.

1905 Einstein publishes his paper 'On the electrodynamics of moving bodies' (Special Relativity Theory).

1906 Schwarzschild attributes energy transport in stellar atmospheres principally to radiation.
Kapteyn lays out his Plan of Selected Areas.
Burnham publishes his double star catalog containing data on 13 665 double star systems.

1907 Emden publishes his book *Gaskugeln* (Gaseous Spheres).
Hertzsprung distinguishes giant and dwarf stars.

1908 Hale discovers the magnetic fields of sunspots.

1910 By Hertzsprung's suggestion Rosenberg succeeds in making a color–magnitude diagram of the Pleiades.
Stebbins successfully experiments with selenium photocells.
According to Schwarzschild's definition of color (1900), the color index (CI) is internationally defined.

1911 Halm publishes the relationship relating the absolute magnitudes and masses of the stars.

1912 Miss Leavitt discovers the period–luminosity relation for Cepheids in the Small Magellanic Cloud.
Hess discovers cosmic rays with a balloon ascent.

1913 For the first time Russell publishes his diagram, which later becomes known as the Hertzsprung–Russell Diagram.

1913 Introduction of photoelectric photometry by Guthnick, Rosenberg and Meyer, and Kunz and Stebbins.

The International Solar Union adopts the Harvard spectral classification.

1914 Adams and Kohlschütter establish the method of spectroscopic parallaxes.

1916 Einstein publishes his Theory of General Relativity.

1917 Einstein develops a static universe model.

Installation of the 100-inch Hooker reflector at Mt Wilson Observatory.

1918 Appearance of the *Geschichte und Literatur des Lichtwechsels der bis Ende 1915 als sicher veränderlich anerkannten Sterne* (History and Literature of Light Variation of Stars known by the End of 1915 to be Definitely Variable).

Shapley discovers the galactic halo.

1919 Founding of the International Astronomical Union.

Brown publishes improved lunar tables.

1920 Saha publishes his theory of ionization in stellar atmospheres.

1922 Friedmann publishes his evolutionary model of the cosmos.

1923 Hubble identifies Cepheids in the Andromeda Nebula and in the Triangulum Nebula and determines the distances to these galaxies.

1926 Eddington publishes his book *The Internal Constitution of the Stars*.

1926/27 Lindblad and Oort develop their dynamical theories of the Milky Way system.

1929 Hubble and Humason discover the relation between the red shifts in the spectra of extragalactic objects and their distances and prove for the first time with sufficient accuracy that the universe is expanding.

1930 Trumpler gives evidence for the existence of generally diffused interstellar matter from investigations of open star clusters.

Lyot develops the coronagraph.

1931 Schmidt builds a telescope with a wide-field coma-free field of view (the Schmidt reflector).

Notes

INTRODUCTION

1. Quoted according to *Johannes Kepler in seinen Briefen* (Kepler's Letters), Max Caspar and Walter von Dyck (eds.), vol. 1, Munich and Berlin, 1930, p. 367.
2. Johann Wolfgang von Goethe, *Materialien zur Geschichte der Farbenlehre* (Notes on the History of Color Theory), *Goethes Werke in fünf Bänden*, vol. 3, Leipzig, 1959, p. 774.
3. Friedrich Engels, *Dialektik der Natur* (Dialectics of Nature), Marx/Engels: *Werke*, vol. 20, Berlin, 1968, p. 313.

I. CONSTRUCTION AND MOTION OF THE HEAVENS – CLASSICAL ASTRONOMY

1. Actually, many astronomers had seen this planet before 1781, without knowing, however, that it was a planet, not a star.
2. Immanuel Kant, *Allgemeine Naturgeschichte und Theorie des Himmels* (General Natural History and Theory of the Heavens; hereafter: Kant, *Natural History*), *Immanuel Kants Werke*, E. Cassirer (ed.), vol. 1, Berlin, 1912, p. 253.
3. Quoted according to Michael A. Hoskin, *William Herschel and the Construction of the Heavens* (New York: Norton, 1964; London, 1963), p. 76.
4. *Ibid.*, p. 174.
5. *Ibid.*, p. 82.
6. Engels, *Dialektik der Natur*, *op. cit.*, p. 314.
7. Kant, *Natural History*, *op. cit.*, p. 231.
8. Engels, *Dialektik der Natur*, *op. cit.*, p. 319.
9. *Ibid.*, p. 316.
10. Kant, *Natural History*, *op. cit.*, p. 316.
11. Friedrich Engels, *Ludwig Feuerbach und der Ausgang der klassischen deutschen Philosophie* (Ludwig Feuerbach and the Emergence of Classical German Philosophy; hereafter: Engels, *Ludwig Feuerbach*), Marx/Engels: *Werke*, vol. 21, Berlin, 1968, p. 295.
12. Wilhelm Olbers, *Drei kosmologische Vorträge*, Stanley L. Jaki (ed.), *Nachrichten der Olbers-Gesellschaft*, no. 79, Bremen, 1970, p. 15.
13. *Georg Christoph Lichtenberg's physikalische und mathematische Schriften*, Ludwig Christian Lichtenberg and Friedrich Kries (eds.), vol. 1, Göttingen, 1803, p. 345.
14. Friedrich Wilhelm Bessel, *Populäre Vorlesungen über wissenschaftliche Gegenstände* (hereafter: Bessel, *Lectures*), H. C. Schumacher (ed.), Hamburg, 1848, p. 17.

15. Later it was proven that this comet had been observed since 240 BC. Its next perihelion passage is expected in February 1986. The comet has already been 'recovered', having been imaged with the 200-inch telescope at Mt Palomar on 16 October 1982.

16. Franz Xaver von Zach, *Monatliche Correspondenz zur Beförderung der Erd- und Himmels-Kunde* (hereafter: *Monatliche Correspondenz*), vol. 5, 1802, p. 394.

17. Arsen Gulyga, *Georg Wilhelm Friedrich Hegel*, Leipzig, 1974, pp. 69–70.

18. Franz Xaver von Zach, *Monatliche Correspondenz*, op cit., p. 334.

19. Friedrich Wilhelm Bessel, Letter to the Royal Academy of Sciences in Berlin of 15 October 1824, *Zentralarchiv der Akademie der Wissenschaften der DDR, Acta der wissenschaftlichen Unternehmungen der mathematischen Klasse*, vol. 2, part II, VII^c, section 2.

20. Bessel, *Lectures*, op. cit., p. 342.

21. Quoted according to Julius Dick, *Die Sterne*, vol. 29, 1953, p. 131.

22. Quoted according to Diedrich Wattenberg, Fr. v. Bessel und die Vorgeschichte der Entdeckung des Neptun, *Mitteilungen der Archenhold-Sternwarte*, no. 54, Berlin-Treptow, 1959, p. 5.

23. Bessel, *Lectures*, op. cit., p. 451.

24. Quoted according to Diedrich Wattenberg, *Johann Gottfried Galle, 1812-1910*, Leipzig, 1963, p. 48.

25. Extract from a letter of Professor C. G. J. Jacobi, Member of the Berlin Academy of Sciences, Berlin, 10 October 1848, *Astronomische Nachrichten*, vol. 28, 1849, cols. 45–46.

26. The result of 9.5 was roughly correct; the modern value is 8″.8.

27. Bessel, *Lectures*, op. cit., p. 242.

28. Friedrich Wilhelm Bessel, Über den Doppel-Stern Nro. 61 Cygni, *Monatliche Correspondenz*, vol. 26, 1812, p. 161.

29. *Monatliche Correspondenz*, vol. 2, 1800, pp. 155–166.

30. *Monatliche Correspondenz*, vol. 16, 1807, pp. 261 ff.

31. The modern value of the oblateness for the international Earth ellipsoid is 1/297.

32. See Dieter B. Herrmann, Unbekannte Briefe von Georg Christoph Lichtenberg an Johann Hieronymus Schroeter, *Vorträge und Schriften der Archenhold-Sternwarte*, no. 21, Berlin-Treptow, 1965.

2. THE ORIGIN OF ASTROPHYSICS

1. Engels, *Ludwig Feuerbach*, op. cit., p. 278.

2. Bessel, *Lectures*, op. cit., pp. 413–414.

3. Carl Friedrich Gauss, Astronomische Antrittsvorlesung, *Mitteilungen der Gauss-Gesellschaft*, no. 7, Göttingen, 1970, p. 27.

4. Johann Heinrich Mädler, *Reden und Abhandlungen über Gegenstände der Himmelskunde*, Berlin, 1870, p. 16.

5. Friedrich Zöllner, *Wissenschaftliche Abhandlungen*, vol. 4, Leipzig, 1881, p. 35.

6. Karl Friedrich Zöllner, *Photometrische Untersuchungen*, Leipzig, 1865, p. 315.

7. Quoted according to Georg Lockmann, *Robert Wilhelm Bunsen*, Stuttgart, 1949, p. 152.

8. Johann Carl Friedrich Zöllner, *Grundzüge einer allgemeinen Photometrie des Himmels*, Berlin, 1861, p. IX.

9. Gustav Müller and Paul Kempf, Photometrische Durchmusterung des nördlichen Himmels, Part 1, *Publikationen des Astrophysikalischen Observatoriums Potsdam*, vol. 9, no. 31, Potsdam, 1894, Introd., p. 5.

10. According to Paul Guthnick, Physik der Fixsterne, *Astronomie*, J. Hartmann (ed.), Leipzig and Berlin, 1921, p. 401.

11. Hermann Carl Vogel, Ueber den Einfluss der Rotation eines Sterns auf sein Spectrum, *Astronomische Nachrichten*, vol. 90, 1877, cols. 74–75.

12. Friedrich Argelander, Aufforderung an Freunde der Astronomie, *Jahrbuch für 1844*, Heinrich Christian Schumacher (ed.), Stuttgart and Tübingen, 1844, p. 227.

13. William Huggins, *Ergebnisse der Spectralanalyse in Anwendung auf die Himmelskörper*, Leipzig, 1869, p. 36.

3. MICROCOSMOS – MACROCOSMOS

1. Arthur S. Eddington, *Stars and Atoms*, Oxford, 1927, p. 10.

2. Max Planck, *Wissenschaftliche Selbstbiographie*, Leipzig, 1948, p. 27.

3. *Ibid.*

4. Friedrich Zöllner, *Wissenschaftliche Abhandlungen*, op. cit., p. 530.

5. Today absolute magnitudes are measured with respect to a unit distance of 10 parsecs (pc). To do this for the data in Table 7, add 5 magnitudes.

6. Karl Schwarzschild, Über das System der Fixsterne, *Himmel und Erde*, vol. 21, 1909, p. 449.

7. R. Emden, *Gaskugeln*, Leipzig and Berlin, 1907, p. III.

8. E. von der Pahlen, *Lehrbuch der Stellarstatistik*, Leipzig, 1937, p. V.

9. Wilhelm Becker, *Sterne and Sternsysteme*, Dresden and Leipzig, 1950, p. 288.

4. TECHNOLOGY AND THE ORGANIZATION OF RESEARCH

1. Dieter B. Herrmann, Zur Statistik von Sternwartengründungen im 19. Jahrhundert, *Die Sterne*, vol. 49, 1973, pp. 48–52. See also by the same author: An exponential law for the establishment of observatories in the nineteenth century, *Journal for the History of Astronomy*, vol. 4, 1973, pp. 57–58, and, Sternwartengründungen, Wissensproduktion, und ökonomischer Fortschritt, *Die Sterne*, vol. 51, 1975, pp. 228–234.

2. *Monatliche Correspondenz*, vol. 21, 1810, p. 46.

3. *Lichtenbergs Briefe*, Albert Leitzmann and Carl Schüdderkopf (eds.), vol. 3, Leipzig, 1904, pp. 206 ff.

4. See Hermann A. Schumacher, Die Lilienthaler Sternwarte, *Abhandlungen des wissenschaftlichen Vereins Bremen*, vol. 11, 1890, p. 78.

AFTERWORD

1. Engels, *Ludwig Feuerbach*, op. cit., p. 267.

Bibliography

The most important individual publications of astronomers during the nineteenth and early twentieth centuries appeared in these journals:

Monatliche Correspondenz zur Beförderung der Erd- und Himmels-Kunde (1800–1813).
Astronomische Nachrichten (since 1821).
Astronomical Journal (1849–1861 and since 1885).
Astrophysical Journal (since 1896).
Zeitschrift für Astrophysik (1930–1966).

Numerous biographies of astronomers, based on the latest historical research, are to be found in:

Dictionary of Scientific Biography, Charles Coulston Gillespie, editor in chief, 16 vols. (New York, Charles Scribner's Sons), 1970–1980.

International research in all areas of the history of astronomy and related sciences is published in:

Journal for the History of Astronomy, M. A. Hoskin (ed.), Cambridge (UK), since 1970.

Compilations of important original works, which are of particular importance in the development of astronomy include:

Astronomy, B. Lovell (ed.), Amsterdam, 1970 (covers research published during the years 1851–1939).
Source Book in Astronomy 1900–1950, H. Shapley (ed.) (Cambridge (USA): Harvard University Press), 1960.
A Source Book in Astronomy and Astrophysics 1900–1975, K. R. Lang and O. Gingerich (eds.) (Cambridge (USA) and London : Harvard University Press), 1979.

A selected bibliography of international research in the history of astronomy is:

David H. De Vorkin, *The History of Modern Astronomy and Astrophysics* (= Bibliographies of the History of Science and Technology, vol. 1), R. Multhauf and E. Wells (eds.) (New York and London: Garland), 1982.

MONOGRAPHS COVERING THE HISTORY OF ASTRONOMY

Abetti, Giorgio, *The History of Astronomy* (London: Henry Schumann), 1952.
Airy, George Biddell, *Abriss einer Geschichte der Astronomie im Anfange des neunzehnten Jahrhunderts (1800–1832)*, C. L. Littrow (transl.), Vienna, 1835.

Becker, Friedrich, *Geschichte der Astronomie*, 3rd ed., Mannheim and Zurich, 1968.

Bernal, John Desmond, *Science in History*, 4 vols., 3rd ed. (Cambridge (USA): MIT Press), 1965.

Bryant, Walter B., *A History of Astronomy*, London, 1907.

Clerke, Agnes M., *A Popular History of Astronomy during the Nineteenth Century*, 4th ed. (London: Adam and Charles Black), 1902.

Doig, Peter, *A Concise History of Astronomy*, London, 1950.

Herrmann, Dieter B., *Entdecker des Himmels*, Cologne, 1979.

Jahn, Gustav Adolf, *Geschichte der Astronomie vom Anfange des 19. Jahrhunderts bis zum Ende des Jahres 1842*, 2 vols., Leipzig, 1844.

Ley, W., *Die Himmelskunde, Eine Geschichte der Astronomie von Babylon bis zum Raumzeitalter*, Düsseldorf and Vienna, 1965.

Mädler, Johann Heinrich, *Geschichte der Himmelskunde der ältesten bis auf die neueste Zeit*, 2 vols., Braunschweig, 1873.

Mineur, Henri, *Histoire de l'astronomie stellaire jusqu'a l'époque contemporaine*, Paris, 1934.

Pannekoek, A., *A History of Astronomy*, London, 1961.

Ronan, Colin A., *Discovering the Universe – A History of Astronomy*, London, 1972.

Struve, Otto and Zebergs, Velta, *Astronomy of the 20th Century* (New York and London: Macmillan), 1962.

Wolf, Rudolf, *Geschichte der Astronomie*, Munich, 1877.

Wolf, Rudolf, *Handbuch der Astronomie, ihrer Geschichte und Literatur*, 2 vols., Zurich, 1890–1893.

Zinner, Ernst, *Die Geschichte der Sternkunde von den ersten Anfängen bis zur Gegenwart*, Berlin, 1931.

LITERATURE FOR THE INDIVIDUAL CHAPTERS

1. *Construction and motion of the heavens – classical astronomy*

Andresen, P. B., *Die Geschichte der Monddistanzen mit besonderer Berücksichtigung ihrer theoretischen und praktischen Grundlagen*, Hamburg, 1924.

Baillaud, Benjamin, *Histoire de l'astronomie de position*, 3 vols., [n. p.], 1933.

Bauschinger, Julius, *Die Bahnbestimmung der Himmelskörper*, 2nd ed., Leipzig, 1928.

Bessel, Friedrich Wilhelm, *Populäre Vorlesungen über wissenschaftliche Gegenstände*, edited after the death of the author by Heinrich Christian Schumacher, Hamburg, 1848.

Chambers, George F., *The Story of the Comets*, Oxford, 1909.

Forbes, Eric G., *The Birth of Navigational Science*, Maritime Monographs and Reports, no. 10, Greenwich, 1974.

Franz, Julius, *Der Mond*, Leipzig, 1906.

Galle, Johann Gottfried, Über die erste Auffing der Planeten Neptun, *Astronomische Nachrichten*, vol. 89, 1877, col. 249, and vol. 101, 1882, col. 219.

Grosser, Morton, *The Discovery of Neptune*, Cambridge (USA), 1962, and New York, 1979.

Hamel, Jürgen, *Friedrich Wilhelm Bessel*, Leipzig, 1984 (in press).

Herrmann, Dieter B., *Komische Weiten: Geschichte der Entfernungsmessung im Weltall* (Leipzig: Johann Ambrosius Barth), 1977; 2nd ed., 1981.

Herschel, William, *Scientific Papers of William Herschel*, J. L. E. Dreyer (ed.), 2 vols., London, 1912.

Hoffmeister, Cuno, *Meteoric Currents*, Weimar, 1948.

Hoskin, M. A., *William Herschel*, New York, 1959.

Hoskin, M. A., *Stellar Astronomy* (Chalfont St Giles (UK): Science History Publications), 1982.

Kant, Immanuel, *Allgemeine Naturgeschichte und Theorie des Himmels*, in *Immanuel Kants Werke*, E. Cassirer (ed.), vol. 1, Berlin, 1912; translated by W. Hastie as *Kant's Cosmogony*, revised and edited by Willy Ley (New York: Greenwood), 1968.

Olbers, Wilhelm, *Drei kosmologische Vorträge*, Stanley L. Jaki (ed.), in *Nachrichten der Olbers-Gesellschaft*, no. 79, Bremen, 1970.

Zach, Franz Xaver von, Über einen zwischen den Planeten Mars und Jupiter längst vermutheten nun wahrscheinlich entdeckten neuen Hauptplaneten unseres Sonnon-Systems, *Monatliche Correspondenz zur Beförderung der Erd- und Himmels-Kunde*, vol. 3, 1801, pp. 592–623.

2. *The origin of astrophysics*

Argelander, Friedrich, Aufforderung an Freunde der Astronomie, in *Jahrbuch für 1844*, Heinrich Christian Schumacher (ed.), Stuttgart and Tübingen, 1844, pp. 222–254.

Armitage, Angus, *A Century of Astronomy*, London, 1950.

Clerke, Agnes M., *Problems in Astrophysics* (London: Adam and Charles Black), 1903.

De Vaucouleurs, Gerard, *Astronomical Photography from the Daguerreotype to the Electron Camera*, London, 1961.

Dormann, L. J., *Cosmic Rays*, New York, 1974.

Gauss, Carl Friedrich, Astronomische Antrittsvorlesung, in Gauss: *Werke*, edited by the Gesellschaft der Wissenschaften zu Göttingen, Göttingen, 1929, pp. 177–199.

Guthnick, Paul, Physik der Fixsterne, in *Astronomie*, J. Hartmann (ed.), Leipzig and Berlin, 1921.

Herrmann, Dieter B., Aus der Entwicklung der Grössenklassen-Definition im 19. Jahrhundert, *Die Sterne*, vol. 48, 1972, pp. 20–30 and 113–120.

Herrmann, Dieter B., Zur Frühentwicklung der Astrophysik in Deutschland und in den USA, in *NTM-Schriftenreihe für Geschichte der Naturwissenschaften, Technik und Medizin*, vol. 10, part 1, 1973, pp. 38–44.

Herrmann, Dieter B. and Hoffmann, D., Astrofotometrie und Lichttechnik in der 2. Hälfte des 19. Jahrhunderts, in *NTM-Schriftenreihe für Geschichte der Naturwissenschaften, Technik und Medizin*, vol. 13, 1976, pp. 94–104.

Jánossy, Lajos, Zum Gedenken an den vor 50 Jahren erbrachten Nachweis der Existenz der kosmischen Strahlung durch V. F. Hess and W. Kolhörster, in *Deutsche Akademie der Wissenschaften, Vorträge und Schriften*, no. 93, Berlin, 1964.

Kedrow, B. M., *Spektralanalyse*, Berlin, 1961.

Lockyer, J. Norman, *The Chemistry of the Sun*, London, 1887.

McGucken, William, *Nineteenth Century Spectroscopy*, Baltimore, 1969.

Mädler, Johann Heinrich, *Reden und Abhandlungen über Gegenstände der Himmelskunde*, Berlin, 1870.

Maunder, E. W., *Sir William Huggins and Spectroscopic Astronomy*, London, 1913; facsimile reprint, London, 1979.

Meadows, A. J., *Early Solar Physics*, Oxford, 1970.

Melnikov, O. A., K istorii razvitiia astrospektroskopii v Rossii i v SSSR, in *Istoriko astronomicheskie issledovaniia*, vol. 3, 1957, pp. 13–258.

Scheiner, Julius, *Populäre Astrophysik*, 2nd ed., Leipzig and Berlin, 1912.

Schellen, H., *Die Spectralanalyse in ihrer Anwendung auf die Stoffe der Erde und die Natur der Himmelskörper*, Braunschweig, 1878.

Schiller, Karl, *Einführung in das Studium der Veränderlichen Sterne*, Leipzig, 1923.

Secchi, Angelo, *Le Soleil*, Paris, 1870.

Seitter, Waltraud, Probleme der Spektralklassifikation, *Mitteilungen der Astronomischen Gesellschaft*, no. 25, 1968, pp. 105–125.

Vogel, Hermann Carl, Ueber den Einfluss der Rotation eines Sterns auf sein Spectrum, *Astronomische Nachrichten*, vol. 90, 1877, cols. 71–76.

Young, Charles, *The Sun*, New York, 1881.

Zöllner, Johann Carl Friedrich, *Grundzüge einer allgemeinen Photometrie des Himmels*, Berlin, 1861.

Zöllner, Karl Friedrich, *Photometrische Untersuchungen*, Leipzig, 1865.

3. Microcosmos – macrocosmos

Berendzen, R., Hart, R. and Seeley, D., *Man Discovers the Galaxies*, New York, 1976.

Bethe, Hans, Energieerzeugung in Sternen, *Die Naturwissenschaften*, vol. 55, 1968, pp. 405–413.

Chapman, S., The source of the Sun's energy, *Monthly Notices of the Royal Astronomical Society*, vol. 102, 1942, pp. 110–130.

Clerke, Agnes M., *The System of the Stars*, London, 1890.

Eddington, Arthur S., *The Internal Constitution of the Stars*, Cambridge (UK), 1926; reprinted New York: Dover, 1959.

Eddington, Arthur S., *Stars and Atoms*, Oxford, 1927.

Einstein, Albert, *Über die spezielle und die allgemeine Relativitätstheorie, gemeinverständlich*, Braunschweig, 1917.

Forbes, Eric Gray, A history of the solar red shift problem, *Annals of Science*, vol. 17, 1961, pp. 129–164.

Grotrian, Walter, Über den Ursprung der Nebellinien, *Die Naturwissenschaften*, vol. 16, 1928, pp. 177–182.

Hermann, Armin, *Frühgeschichte der Quantentheorie (1899–1913)*, Mosbach, 1969.

Herneck, Friedrich, *Bahnbrecher des Atomzeitalters*, 7th ed., Berlin, 1975.

Herrmann, Dieter B., Gedanken zur Sternentwicklung in der Anfangszeit der Spektroskopie, *Wissenschaft und Fortschritt*, vol. 24, 1974, pp. 537–541.

Hertzsprung, Ejnar, Zur Strahlung der Sterne (= *Ostwalds Klassiker der exakten Wissenschaften*, vol. 255), D. B. Herrmann (ed.), Leipzig, 1976.

Hindmarsh, W. R., *Atomspektren* (vol. 76 in a series of scientific paperback texts), Berlin–Oxford–Braunschweig, 1972.

Hoskin, Michael A., Apparatus and ideas in mid-nineteenth century cosmology, *Vistas in Astronomy*, vol. 9, 1968, pp. 79–85.

Klauder, H., Das grössere Milchstrassensystem, *Die Sterne*, vol. 25, 1949, pp. 106–111.

Kobold, Hermann, Der Bau des Fixsternsystems, *Die Wissenschaft*, no. 11, Braunschweig, 1906.

Meurers, J., Die Entwicklung der Theorie des inneren Aufbaus der Sterne, *Die Sterne*, vol. 29, 1953, p. 465.

Nielsen, Axel V., Contributions to the history of the Hertzsprung–Russell Diagram, *Centaurus*, vol. 9, 1963, pp. 219–253.

Oort, Jan H., The development of our insight into the structure of the galaxy between 1920 and 1940, *Education in and History of Modern Astronomy, Annals of the New York Academy of Sciences*, vol. 198, 1972, pp. 255–266.

Pahlen, E. von der, *Lehrbuch der Stellarstatistik*, Leipzig, 1937.

Seeley, D. and Berendzen, R., The development of research in interstellar absorption, 1900–1930, *Journal for the History of Astronomy*, vol. 3, 1972, pp. 52–64 and 75–86.

Waterfield, R. I., The story of the Hertzsprung–Russell Diagram, *Journal of the British Astronomical Association*, vol. 67, 1956, pp. 2–24.

4. *Technology and the organization of research*

Ambronn, L., *Handbuch der astronomischen Instrumentenkunde*, 2 vols., Berlin, 1899.

Brush, Stephen G., The rise of astronomy in America, *American Studies*, vol. 20, 1979, pp. 41–67.

Esche, G. and Kessler, H. E., *Der VEB Carl Zeiss Jena – Seine äussere Entwicklung von der Gründung bis zur Gegenwart, Schriftenreihe des Stadtmuseums Jena*, 1966.

Glaze, Francis W., The optical glass industry, past and present, *Annual Report of the Smithsonian Institution*, [Washington, DC], 1948, pp. 217–225.

Herrmann, Dieter B., Das Astronomentreffen im Jahre 1798 auf dem Seeberg bei Gotha, *Archive for the History of Exact Sciences*, vol. 6, 1970, pp. 326–344.

Herrman, Dieter B., Gould and his *Astronomical Journal, Journal for the History of Astronomy*, vol. 2, 1971, pp. 98–108.

Herrmann, Dieter B., Die Entstehung der astronomischen Fachzeitschriften in Deutschland (1798–1821), *Veröffentlichungen der Archenhold-Sternwarte*, no. 5, Berlin-Treptow, 1972.

Herrmann, Dieter B., An exponential law for the establishment of observatories in the nineteenth century, *Journal for the History of Astronomy*, vol. 4, 1973, pp. 57–58.

Herrmann, Dieter B., Sternwartengründungen, Wissensproduktion und ökonomischer Fortschritt, *Die Sterne*, vol. 51, 1975, pp. 228–234.

Herrmann, Dieter B., Zur Vorgeschichte des Astrophysikalischen Observatorium Potsdam, *Astronomische Nachrichten*, vol. 296, 1975, pp. 245–259.

Jebsen-Marwedel, H., Utzschneider, Reichenbach, Fraunhofer, Zeiss, Abbe, Schott: Zwei Triumvirate an der Wiege wissenschaftlich fundierter Glastechnik, *Glastechnische Berichte*, vol. 39, 1966, pp. 334–339.

King, Henry C., *The History of the Telescope*, London, 1955.

Krisciunas, Kevin, A short history of Pulkovo Observatory, *Vistas in Astronomy*, vol. 22, part 1, 1978, pp. 27–37.

Learner, Richard, *Astronomy through the Telescope* (New York: Van Nostrand Reinhold), 1981.

Meyer, F., Über die Entwicklung der astronomischen Instrumente im Zeisswerk Jena, *Zeitschrift für Instrumentenkunde*, vol. 50, 1930.

Mikhaïlov, A. A. (ed.), *Astronomiia v SSSR za sorok let 1917–1957*, Moscow, 1960.

Musto, David F., The development of American astronomy during the early nineteenth century, *Ithaca*, 26 VIII–2 IX, 1962, pp. 733–736.

Repsold, J. A., *Zur Geschichte der astronomischen Messwerkzeuge*, 2 vols., Leipzig, 1908–1914.

Riekher, R., *Fernrohre und ihre Meister*, Berlin, 1957.

Roth, Günter D., Entwicklung der optischen Industrie in München im 19. Jahrhundert, *Technikgeschichte in Einzeldarstellungen*, no. 19, Düsseldorf, 1971, pp. 187–212.

Schwarzschild, Karl, Präzisionstechnik und astronomische Forschung, *Deutsche Mechaniker-Zeitung*, 1914, part 14, pp. 149–153, and part 15, pp. 162–165.

Seitz, A., *Joseph Fraunhofer und sein optisches Institut*, Berlin, 1926.

Warner, Deborah J., Alvan Clark & Sons: artists in optics, *Bulletin 274 of the National Museum of the Smithsonian Institution*, Washington, DC, 1968.

Woronzow-Weljaminow, Boris A., Die Entwicklung der Wissenschaften in der UdSSR-Astronomie, *Enzyklopädie der Union der Sozialistischen Sowjetrepubliken*, vol. 2, Berlin, 1950, pp. 1369–1378.

Name index

Abbe, Ernst (1840–1905) 167
Abney, William de Wiveleslie (1843–1920) 96
Adams, John Couch (1819–1892) 36–7, 62, 199
Adams, John Quincy (1767–1848) 182
Adams, Walter Sydney (1876–1956) 48, 111, 126, 201
Airy, George Biddell (1801–1892) 32, 37
Alfvén, Hannes (1908–   ) 107
Ångström, Anders Jonas (1814–1874) 99
Arago, Dominique Francois Jean (1786–1853) 81, 85, 112, 193, 199
Argelander, Friedrich Wilhelm August (1799–1875) 32–33, 71, 88, 90, 108, 147
Aristotle (384–322 BC) 3, 15
Auwers, Arthur von (1838–1915) 34–35, 44

Baade, Walter (1893–1960) 141
Barnard, Edward Emerson (1857–1923) 44, 62, 85, 142
Bayer, J. J. 52–53
Becker, W. 94, 123
Beer, Wilhelm (1797–1850) 57, 65, 67
Belopol'skiï, Aristarkh Apollonovich (1854–1934) 96, 111
Benzenberg, Johann Friedrich (1777–1846) 198
Bernoulli, Daniel (1700–1782) 34
Bessel, Friedrich Wilhelm (1784–1846) 20, 44, 50, 55, 67, 69, 112, 163
    compilation of star catalogs 30–2, 34–6
    61 Cygni parallax 42, 46–7, 199
    Fundamenta Astronomiae 34, 189, 198
    geodetic measurements 52–3
    parallax measurements 42, 45–7
Bethe, Hans (1906–   ) 196
Bode, Johann Elert (1747–1826) 24, 188, 190, 193
Bohlin, Karl Petrus Theodor (1860–1939) 148
Bohnenberger, Johann Gottlieb Friedrich von (1765–1831) 189, 198

Bohr, Niels Henrik David (1885–1962) 117–19
Boltzmann, Ludwig (1844–1906) 98, 115, 117
Bonaparte, Napoleon (1769–1821) 21n
Bond, George Phillips (1825–1865) 60, 75
Bond, William Cranch (1789–1859) 60, 62, 83, 84
Borda, Jean-Charles (1733–1799) 52
Boss, Lewis (1846–1912) 48, 147
Bothe, Walther Wilhelm Georg (1891–1957) 135
Bouvard, Alexis (1767–1843) 36
Bowen, Ira Sprague (1898–1973) 120
Bradley, James (1693–1762) 34, 36, 42, 45, 50
Brahe, Tycho (1546–1601) 2, 15, 18n, 110
Brandes, Heinrich Wilhelm (1777–1834) 198
Bredikhin, Fëdor Aleksandrovich (1831–1904) 59, 112
Brewster, David (1781–1868) 99
Brill, Alfred Otto (1885–1949) 122
Brown, Ernest William (1866–1938) 50, 201
Bruno, Giordano (1548–1600) 4
Buffham, W. 61
Buffon, Georges-Louis Leclerc, Comte de (1707–1788) 15
Bunsen, Robert Wilhelm Eberhard (1811–1899) 77, 80, 118, 199
Burnham, Sherburne Wesley (1838–1921) 44n, 200

Campbell (Admiral) 50
Cannon, Annie Jump (1863–1941) 91, 93–4, 200
Carpenter, J. 68n
Carrington, Richard Christopher (1826–1875) 32
Carrochez, N. S. 159
Cassini, Gian Domenico (1625–1712) 58–60, 62
Catherine II (Russia) (1727–1796) 39
Cauchoix, R. A. 161
Celoria, Giovanni (1842–1920) 136

Subject index

217